T0320176

ROUTLEDGE LIBRARY EDITIONS:
FORESTRY

Volume 3

STOVES AND TREES

STOVES AND TREES

How Much Wood
Would a Woodstove Save
If a Woodstove
Could Save Wood?

GERALD FOLEY,
PATRICIA MOSS
AND
LLOYD TIMBERLAKE

LONDON AND NEW YORK

First published in 1984 by Earthscan.

This edition first published in 2024
by Routledge
4 Park Square, Milton Park, Abingdon, Oxon OX14 4RN

and by Routledge
605 Third Avenue, New York, NY 10158

Routledge is an imprint of the Taylor & Francis Group, an informa business

British Library Cataloguing in Publication Data
A catalogue record for this book is available from the British Library

ISBN: 978-1-032-77116-8 (Set)
ISBN: 978-1-032-76730-7 (Volume 3) (hbk)
ISBN: 978-1-032-76736-9 (Volume 3) (pbk)
ISBN: 978-1-003-47984-0 (Volume 3) (ebk)

DOI: 10.4324/9781003479840

Publisher's Note
The publisher has gone to great lengths to ensure the quality of this reprint but points out that some imperfections in the original copies may be apparent.

Disclaimer
The publisher has made every effort to trace copyright holders and would welcome correspondence from those they have been unable to trace.

STOVES AND TREES

how much wood
would a woodstove save
if a woodstove
could save wood?

Gerald Foley
Patricia Moss
Lloyd Timberlake

 An Earthscan Paperback

© Earthscan 1984
ISBN No 0-905347-51-X

Published by the International Institute for
Environment and Development, London and
Washington, DC
Printed by Russell Press, Nottingham, UK
Typeset by Kerrypress, Luton, UK

This book was edited and produced by
Jon Tinker, Geoffrey Barnard and
Barbara Cheney.

Stoves and trees was researched as part of
a project funded by the Swedish
International Development Authority,
and published with funding from the
Netherlands Foreign Ministry. But this
book should not be taken to reflect the
views of these or any other organisations
which support Earthscan. It has drawn
upon the Earthscan Energy Information
Programme's Technical Report No2:
*Improved Cooking Stoves in Developing
Countries.*

Cover photo: Mark Edwards/Earthscan

 Contents

 **Summary**

Chapter 1
The experts claim that widespread use of improved woodburning stoves in the Third World could save wood and slow deforestation. But clearing for farmland, not for fuelwood, is more often the driving force behind deforestation. And because stoves are inefficiently used and deteriorate, because wood is burnt for reasons other than cooking, and because improved stoves cannot be got to everyone, national wood savings through stove programmes can never be significant. However, though improved stoves may not save trees or forests, they can improve the daily lives of human beings.

Chapter 2
The open, "three-stone" fire — used throughout the Third World — is dirty and dangerous. But, unlike most stoves, it is portable and easy to regulate; it burns fuel of all types, shapes and sizes; and it provides light, heat and a social focus. It drys and cures crops stored above it, and it keeps insects out of thatched roofs. Laboratory tests have found that well-tended fires shielded from wind are surprisingly efficient.

Chapter 3
Traditional stoves exist in a variety of forms throughout the Third World, ranging from the hole-in-the-ground *chula* of Bangladesh to the complex stoves which are integral parts of Korean and Chinese homes. They have evolved to fit neatly with local foods, cooking habits, fuels and housing. They are of local materials and are built and repaired locally.

Chapter 4
About half the world's people cook with wood. In some places it is freely gathered; in others it is an increasingly expensive commodity, a condition which tends to reduce demand and encourage conservation. Dung supplements wood on the Indian subcontinent and in parts of southern Africa and South America. Many people want to move directly from wood to kerosene, gas or electricity. Energy consumption varies widely: from country to country but also among families in the same village. Where fuel is plentiful, people use a lot of it; where it is scarce, they use less of it, whatever their cooking method.

Chapter 5
Designers of new stoves must balance cheapness of materials and simplicity of manufacture against durability; efficiency against ease of use. Options include improving the open fire by shielding it, converting buckets into simple charcoal stoves, or manufacturing more complex stoves. Mud stoves are cheap but deteriorate rapidly. Draught controls improve efficiency but tend to break down; chimneys remove smoke but add to cost and complexity. Thermal efficiency and performance are both hard to measure, especially in Third World villages.

Chapter 6
The first stove programmes (1950s to 1960s) in India and Indonesia were meant to save wood and free women from "smoke, soot, excess heat, wasted fuel and fire risk". But the early efforts had little long-term effect. Sophisticated stove programmes now exist in Guatemala, Senegal, Upper Volta, Kenya, Indonesia, Sri Lanka, Nepal, Niger, India and elsewhere.

Chapter 7
Few stove programmes have devoted the time or money necessary to find out how much wood their stoves save. The few studies made show that some new stoves save fuel while others are no more efficient than traditional stoves. Almost no surveys have been made after the novelty of the new stove has worn off and it may have begun to deteriorate.

Chapter 8
Stove programmes compete for development aid against efforts to improve health, education and agriculture. Only programmes which can ultimately become independent and self-sustainable can be justified, as they are the only ones which can ever reach many people. Stove experts have come up with nine steps toward disseminating stoves on a large scale. Commercially viable stoves are the easiest to disseminate; but even with these, marketing work is required. Low-cost or no-cost stoves are often of poor quality. Direct subsidies may help initially, but there must be a mechanism for phasing these out if the programme is to have any long-term impact.

Introduction

It is a widely held belief that if only the people in developing countries who cook on open fires would switch to new, improved wood and charcoal burning stoves, then the planet's forests would be saved. "To solve the problem of deforestation we should have 100 million stoves within 20 years", wrote a stove expert in 1982.

Laboratory tests prove that — in laboratories — well-designed stoves burn much less wood than open fires. So in the past decade a great deal of time, energy and money has been spent by aid agencies and Third World governments to design better stoves and get them into Third World dwellings.

This book examines such "stove programmes" in Africa, Asia and Latin America, and analyses how more traditional stoves, open fires and domestic fuels are used in the Third World, as well as some of the driving forces behind deforestation. It concludes that much of the time and money invested in stove programmes to date has been wasted.

Many people in developing countries want and need improved stoves. It may even prove possible for aid agencies, governments and voluntary organisations to learn from past mistakes and find efficient means of getting efficient stoves spread widely throughout a society. But it is a harder task than the experts previously believed.

In most of the world, the constant search for fuelwood by the two billion people who rely on wood as their main energy source is not the main reason why forests are being destroyed. In places where populations are growing and where there are still forests — conditions true in much of the tropical Third World — the clearing of land for agriculture can consume 100 times more trees than the search for fuelwood. In fact, clearing for agriculture can produce a surplus of fuelwood which gives people little incentive or desire to conserve wood. Where both forests and wood are scarce, the poor may switch to dung, crop residues and other "free" fuels, while the rich change to kerosene or gas. Neither group is in the market for new woodburning stoves; the poor cannot afford them, and the rich want something better.

Stoves may in laboratory tests save 50% of the wood which would have been used to accomplish the same task over an open fire. But by the time allowances are made for inefficient use of the stoves in the villages; for stove deterioration; for jobs stoves cannot do such as heating, lighting, or food drying; for the other uses of wood; then a "successful" stove dissemination programme may wind up saving a nation only 1.5% of the wood it normally uses.

Open fires can be dirty, unpleasant and dangerous — especially for children. But they have some advantages over stoves. They can be built anywhere. Shielded from wind and nursed with care, they are more efficient than generally thought. Fuel of all shapes and sizes can be used. They provide light, heat and a social focal point for family and friends. Their smoke, though a nuisance and a health hazard, keeps insects out of a thatched roof and may be used to preserve foods. Fires — and their fuel when it is gathered — are free. Thus it is often difficult to persuade families that a new stove is better than the fire which has provided for previous generations.

Improved stove designers are also competing with the countless models of traditional stoves spread throughout the developing world. Though they may not be as fuel-efficient as some new designs, most have evolved to mesh well with their users' main foods, cooking utensils, housing and climate.

Domestic fuels move in complex patterns in the Third World. In a single Indian village, near neighbours may rely on different energy sources: electricity, kerosene, charcoal, purchased wood, gathered wood, dung and crop residues. In some areas fuels are seasonal, with villages in parts of southern Africa using dung in winter, wood in summer. Many people who want to "move up" from open fires want kerosene or gas, not a stove which burns wood.

Measuring the amount of fuel families actually use is difficult and costly, and few reliable studies exist. How much of the wood on a fire is being used for cooking, how much for heat, and how much for light? When and why do people change fuels? One of the few reliable conclusions thrown up by both local and regional studies is that the amount of fuel people burn depends to a great extent on what is available to them. If firewood is scarce, they conserve it; if it is plentiful, they tend to burn it — no matter how thermodynamically efficient their stoves or cooking methods may be.

To come up with acceptable stoves, designers must take into account not only technical performance, but the expense and availability of the materials from which they are made, the complexity of the manufacturing process, cost to the consumer, skill required to operate the stoves effectively and local cooking needs and traditions. Many "improved stove" programmes fail because they have not correctly balanced these variables.

During the 1950s and 1960s, stove programmes spread across India as part of the rural development work inspired by Gandhi. They were meant to save wood, but also to liberate women from excess smoke, soot, heat and fire hazards. Despite these efforts — and similar work in Indonesia and Ghana — the early stove programmes had little impact on Third World villagers, but a great impact on the stove programmes which followed. Today, improved stove programmes are operating in Guatemala, Senegal, Upper Volta, Kenya, Indonesia, Sri Lanka, Nepal, Niger, India, and other countries in Africa, Asia and Latin America.

But do these improved stoves save wood in the villages? Strangely, many stove enthusiasts are more concerned with counting the stoves they have "disseminated" than with answering the basic question of whether they are saving fuel. Few follow-up studies meet the requirements of basic statistical analysis. Some of the surveys found energy savings of 20-35% compared to open fires. But other studies showed little savings, and there have been few studies of family fuel use after new stoves have been in place for several months and initial enthusiasm has worn off. According to one expert, "In Upper Volta, two weeks of daily measurements of wood consumption in households *without* stoves was alone sufficient to reduce consumption by 25%".

These studies of stoves in actual use have thrown up few hard conclusions. Some new stoves save wood initially; some do not. In most cases, little is known about what actually happens, especially over a period of months. But there is no worldwide justification for the often repeated claim that stoves save 50% of the wood normally used.

If stove programmes are worth doing at all, then they are worth doing better. Stove experts agreed at a 1981 workshop in Sri Lanka on a nine-step approach to the dissemination of new stoves. Many of the programmes which have failed have either left out some of these steps or taken them in the wrong order. Are the stoves to be sold? If so, their advocates will need to investigate their potential in the marketplace, just as any commercial company wishing to launch a new product would do. Are they to be given away? Then how can the project leaders be sure the stoves will continue to be produced and distributed when the project money runs out? Does one subsidise the stoves heavily, so that many will be distributed quickly, or sell them at closer to their market value? How does one discontinue the subsidy while continuing the dissemination programme?

An improved woodburning stove can make a dwelling cleaner, safer and more pleasant to live in. It can even, if properly used and kept in good repair, save fuel. But stoves are unlikely to slow deforestation or even to save fuel on a national level. Are they the appropriate work of aid agencies? According to one expert:

> "If significant wood savings are not a serious prospect, stoves have to be considered more carefully alongside the whole range of ways in which domestic life can be improved: eg, by training in health or nutrition, inclusion of women in agricultural extension programmes, provision of potable water, improved schools and clinics, etc, etc. When stoves are viewed in this way in relation to the alternative ways of helping women — both women and governments are almost certain to assign them a very low priority."

It is a harsh judgement. But it is one which many experts are prepared to accept as valid in relation to much of the work which has gone on in the past. If stove programmes are to be a continuing focus of aid and technical assistance, the questions and doubts which have now become too urgent to ignore will have to be answered.

A stove programme may not be the absolute optimum investment of resources; but then few projects of any kind can stand that kind of scrutiny. The basic requirement is that any stove programme should be able to provide a rational justification for itself and for the money invested in it. Only those schemes which can do so, and which show signs of ultimately becoming independent and self-sustaining, can warrant the investment of time and resources involved.

Experts such as Professor Phil O'Keefe, head of energy and development research at the Beijer Institute of the Royal Swedish Academy of Sciences, believe that stove programmes do have a role in development efforts:

> "There is a limited, but exciting, opportunity for stove development. It will be urban based and probably involve some degree of mass production. Cutting urban consumption, especially of charcoal, will have a substantial impact on rates of deforestation and will free wood resources for other high technology uses".

He and others feel confident that new stove programmes now being formulated — what designers are calling the "third generation" stove programmes — will more than adequately rise to the challenge. This report indicates just how stiff that challenge is.

Chapter 1

Improved stove programmes: why?

A reason often given for embarking on improved cooking stove programmes is that they will save large quantities of fuelwood and charcoal at a national level. It has been said that if every Senegalese family cooked on an improved stove there would be a 50-60% saving in wood, and the entire forestry deficit of the country would be absorbed. In India, a recent publication says: "Just by doubling the energy efficiency of woodstoves, which is a miserable 5-10% at the moment, the country could halve its present firewood consumption to about 75 million tonnes".

There is no denying the social and environmental problems which come from deforestation. An increasing scarcity of woodfuel forces people to spend more time and money on obtaining the domestic fuel they need. This can bear particularly heavily on the very poor. Collecting wood may divert labour away from growing crops or earning money. Children may have to collect fuel at the expense of their education.

Environmentally, the effects of deforestation can also be very damaging. The loss of trees, which protect the ground against the impact of rain, and of roots, which bind the soil, may speed the erosion of topsoil. In dry areas the disappearance of trees can lead to dust storms and wind erosion. The catalogue of the ill effects of deforestation is becoming all too familiar. There are no doubts about the gravity of deforestation and its effects; but there are growing doubts about the assumed linkage between deforestation and consumption of wood for domestic fuel.

What causes deforestation?

It is often assumed that trees are disappearing because people are using too much fuelwood; the presumption is that if they were to use less fuelwood, the trees would remain. The fallacy in this argument lies in the fact that it ignores the root causes of deforestation.

The most common reason is land hunger. Expanding populations need land on which they can grow food to keep themselves alive. In all the countries with major fuelwood problems, there is also intense population pressure on the available land. Forests represent an unused agricultural potential which is often irresistably attractive to both commercial food producers and to land hungry peasants. Vast forest tracts are cleared for cattle ranching in some of the Latin American countries; peasant families are illegally farming official forest lands all over the developing world.

Ironically, forest clearance may, for the short term, produce a surplus rather than a scarcity of wood. Peasants or ranchers taking over forest land frequently find the standing stock of trees is vastly greater than their fuel demands. They usually burn up the wood just to dispose of it. But in some cases it may enter the fuelwood market. Much of the present fuelwood supply for Managua, Nicaragua, is provided by wood cleared from lands being redistributed under the land reform legislation of the Sandinista government.

The relative impact of land clearance and gathering for fuelwood can be dramatically illustrated by considering the effects of both in an area where there was previously an equilibrium between supply and demand — an equilibrium which has now broken down under the pressure of increasing population. Suppose, for the sake of the example, that the population growth is 3% per annum, the average family land holding is three hectares and fuel consumption three tonnes per year, and the total biomass stock in the forest is 300 tonnes per hectare, a figure typical of many types of temperate forests. Under such circumstances, if the agricultural frontier expands into the forest in order to keep pace with population growth, the loss of wooded land as a result will be exactly 100 times greater than that required to supply the additional woodfuel requirements.

In such circumstances, the wood available for fuel will obviously become scarcer with the passage of time, as the area from which it can be collected diminishes. People will be forced to walk further, take poorer quality wood, and switch to other fuels such as dung or kerosene. The use of a more efficient stove would reduce the amount of fuel they needed to meet their cooking and other requirements, or it would help them make better use of the fuel they are able to obtain. But it would have little, if any, effect on the rate of deforestation occurring as a result of clearing the land for agriculture.

The example arbitrarily concentrates the two effects of fuel gathering and forest clearing into a small area for the sake of illustrating the point. In real life, there will rarely be such a clear-cut situation. But the underlying reality of land hunger will nevertheless remain as one of the most potent agents in causing the depletion of forests.

In arid areas where crop production is difficult, if not impossible, the pressure on wood resources tends to come from grazing animals. As the human population grows, so does that of animals. The open forests on which people depend for fuelwood are increasingly threatened. Browsing animals which previously did little or no damage to a tree now take so much from it that they leave it stunted or dead. Above all, forest regeneration is completely stopped because the seedlings of new trees are eaten as soon as they sprout. Again, improved stoves may help people deal better with their deteriorating fuel position; but they will not significantly alter the process of forest destruction.

Urban woodfuel·demands

In many developing countries, much of the total national fuelwood demand occurs in the cities, sometimes in the form of charcoal.

The impact of the urban fuelwood market is generally felt close to the cities. It may, however, also be felt at considerable distances away if there is an effective road transport system for wood. In Senegal, for example, much of the charcoal for Dakar comes from the Casamance region in the very south of the country, a road journey of around 300 km (190 mi).

The supply of the urban fuel market is often the work of the poor in the rural areas. Trees are cut either for making charcoal, or for direct sale to dealers who come out from the cities. The sight of bundles of fuelwood or bags of charcoal waiting for collection by the roadside is a familiar one in many developing countries. Frequently, much of the wood supply is stolen from reserved forest lands, which are thus subject to completely uncontrolled depletion.

In West Bengal, India, there are areas which were covered with rich forests within the last twenty years. Now, under the onslaught of illegal cutting, these forests have been reduced to metre-high (three-foot) scrub which extends to the horizon.

It has been suggested that the introduction of improved cooking stoves, whether for wood or charcoal, in the urban areas might be able to reduce such impacts. The exact effect, however, of the successful introduction of more efficient stoves in urban areas is difficult to predict. The immediate result of a drop in the use of charcoal for domestic purposes would most probably be a fall in its price. This in turn would tend to tempt other users, whether domestic, commercial, or industrial, into using charcoal, partially cancelling the original reduction in consumption.

What would be the effects on the suppliers in the rural areas? Reduced prices would be less likely to deter the very poor who depend upon illegal cutting than they would producers acting under forest department controls. There is no guarantee that in some cases the level of unauthorised destruction of forest lands might not become even worse as people strove to keep their income from wood sales up.

The position as far as stoves for urban use is concerned is thus similar to that of stoves for rural use. There is no doubt that if they are properly designed and used they could save fuel for individual families. But what, if any, effects this would have on either forests or the fuelwood market remains unknown.

Are there national savings?

Another way of looking at the possible national energy savings is to begin not with the forces causing depletion of forests, but with the stove programme itself. Adding all the likely individual savings which an effective programme would bring, would there be significant savings at a national level?

US energy economist David French, now with the Malawi Ministry of Agriculture, has been working with stoves and studying energy uses in Malawi. His rather pessimistic analysis shows how highly impressive theoretical or laboratory savings by improved stoves are gradually whittled away by practical realities as these stoves are spread throughout the villages.

In laboratory tests in which standard meals were cooked, the *Malawi* stove, a mud stove, produced a 50% wood saving compared to an open fire. These laboratory savings fell by half when standard meal tests were carried out on newly built stoves in villages. The maximum reduction in fuel consumption in the field was 25%. Moreover, mud stoves deteriorate rapidly, causing a fall of 33% in their initial efficiency.

Cooking represents about 60% of domestic fuel use in Malawi. The rest is for heating, lighting, fish and meat drying, beer brewing, large-scale water heating and other cooking tasks which would not be carried out on a new mud stove. Thus, even if every family had a stove, these other uses would consume the same amount of wood. Also, there are times when the weather is too cold or too hot for cooking indoors on a mud stove. There are times when maize cobs or other fuels unsuitable for the stove are commonly used.

Taking all this into consideration, the average long-term savings by a family using the mud stove amount to only about 7.5%. If half the rural population could be supplied with stoves through a massive dissemination campaign extending over a period of years, the savings would amount to 3.8% of the wood used by rural households. Since only 40% of the national wood demand is attributable to rural households in Malawi, the stove programme would save the country a mere 1.5% of its total annual wood use.

The details of the maths are open to question. In particular, the field test results and the longer term behaviour of some ceramic or other stoves might be better than mud stoves. But the point remains that even under optimistic assumptions it is unlikely that the total national savings from a huge and very successful stove diffusion programme would exceed 5%.

Why bother then?

The fact that improved stove programmes are unlikely to prevent deforestation or reduce the national level of energy consumption may be a disappointment to many. But it is improbable that many people are going to change their cooking habits and adopt new stoves for the, to them, abstract goal of lowering national wood consumption, however desirable that goal might appear to the outside observer. Some stove programmes have failed partly because they focussed on such lofty objectives rather than the problem of easing the very real and heavy burdens of domestic existence in so many parts of the developing world.

Experienced stove workers have always known that programmes succeed or fail depending on how potential users rate the merits of the stoves being promoted. Where people see benefits to be obtained from a new stove they will tend to adopt it, provided these benefits are not outweighed by the costs. This is the crucial test which decides whether a stove programme works or not.

The benefits obtained may be in actual energy savings, or in a more effective use of the fuel available. But stoves may also be valued because they make cooking quicker, safer, and cleaner. They protect children from the danger of burns from the open fire. They can reduce respiratory diseases, and ease the burdens of keeping houses, furniture, and clothing clean and free of soot and tar from an open fire. Properly designed stoves, adapted to their environment, and meeting the needs of their users have been a mark of progress and rising quality of life through human history since the beginnings of settled habitation.

The future justification for stove programmes has to be found in their capacity to deliver such direct benefits to the people who will use them. If they do not, then there really is no point in bothering.

Chapter 2

The open fire: pros and cons

Most poor, rural Third Worlders cook on an open fire. Anyone wishing to understand how and why people may want to switch to stoves, must first understand how and why people cook over fires, as well as what advantages fires have over stoves.

Most cooking fires are surrounded by three or more stones, bricks, mounds of mud or lumps of other fireproof material — thus the common name of "three-stone fire". The stones support the cooking pot over the centre of the fire.

In some areas, the stones are arranged so that the pot sits partly down between them; in other places, earth is scooped out between the stones so that the fire is slightly sunk into the ground. In some, the stones are positioned so that the pot can be set back from the fire for slow boiling or simmering.

In Ghana, the three-stone fire is called the *swish stove*. It generally has three mud mounds as its pot supports, but other materials such as concrete blocks, stones, inverted clay pots covered with mud, old car wheels or scrap metal are also used. Women cook sitting on a small stool beside the fire. The pot supports of Upper Volta fires are made of similar materials, and a bride's mother may install the inverted clay pots in her daughter's new home, having filled them with special charms. Sometimes five stones are used to accommodate two fires and two pots.

In Latin America, the open fire is often on a raised adobe or mud platform called a *poyo*; or it may be on a barrel or wooden framework filled with earth. In the rural areas of Guatemala, for example, about 80% of the households rely solely on an open fire for their cooking. Of this total, about 75% are raised, with the rest on the floor.

Some 92% of rural households in Fiji do some or all of their cooking over an open fire. Around 72% of those who do all their cooking over an open fire do so in the traditional Fijian style and cook with the fire on the ground. Over some fires are wooden racks holding firewood and other materials to be dried.

The remaining 28% of Fijian households use the Indian method and cook over a fire on a hearth raised around 60-90 cm (24-35 inches) above the floor. These raised open fires are often built under a broad corrugated iron chimney which funnels the smoke out from the kitchen. This is known as the *Indian fireplace*.

Many people who cook over open fires use a metal trivet: a horizontal metal ring to which three legs are attached. These can hold a single pot over an open

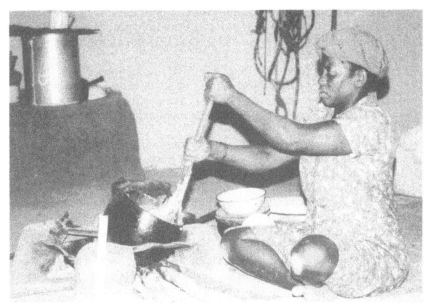

Jas Gill

Cooking "sadza" (maize meal) in Zimbabwe means prolonged stirring over the fire. In this "four-stone fire" iron bars embedded in mounds of clay support the pot and prevent it from sliding about.

fire. In Senegal, 95% of the rural population cook over open fires, with 40% using the three-stone fire, and the other 55% using trivets. The trivet is also used in Upper Volta and is commonly found in Nepal. A rectangular version, in the form of a horizontal four-legged iron frame into which a number of open pot-holders are welded, is becoming increasingly popular in parts of Zimbabwe.

In Zimbabwe, people also use a four-stone fire: four mounds of clay supporting iron bars forming a square, with two diagonal bars added for pot supports. Though they consume more fuel than the traditional three-stone fire, both the iron frame and the four-stone fire are becoming increasingly popular. They allow quicker cooking and are regarded as more modern. In Ecuador, iron bars spanning the fire and resting on stones or bricks on either side are sometimes used to support the cooking pot.

Many African women cook outdoors during the dry season and indoors during the wet season. In such households there are usually two hearths. For these women, cooking outside, even in hot weather, is preferable to the smoky conditions caused by cooking indoors.

In Ghana, Kenya and Senegal, cooking is frequently done inside a special cooking hut, with the fire by an opening in the wall so at least some smoke can escape. The hut is still smoky, but provides protection against wind, rain and dust. Many parents also prefer a cooking hut to keep the fire away from young children.

Where men take more than one wife, wives sometimes share cooking facilities; elsewhere separate families may share facilities. Around 30% of rural families in Ghana share a room for cooking with other households. In Burundi, families often cook within a common courtyard, allowing neighbours to chat while cooking.

In the Sudan, women often cook in the houses of others. It is common for women to spend much of the day when their husbands are at work in their mother's house. They cook their own family's food there and bring it home at the time of their husband's return from work.

Cooking pots may be of clay, aluminium or iron, depending on the family's income, the type of fuel used and the food being cooked. In Nepal, copper cooking pots were traditionally made from locally mined copper and, although mining has now ceased, many are still used. They are repaired and remade by local smiths. Gourds are used as lids for clay cooking pots in some African countries. Flat metal plates are used to cook the various unleavened pancake types of bread found in different regions, such as the Latin American tortillas and the Sudanese kisera.

Advantages of the open fire

The open fire possesses important advantages. It costs nothing, and no special materials, tools or skills are needed to construct it. It can be put wherever is most convenient, and can be moved easily and as often as desired.

The cook can also relatively easily control the heat output from the fire in order to meet the varying needs of cooking. Wood — either small pieces or long sticks which protrude out along the ground from the fire — is pushed in from the sides through the openings between the stones. The size and temperature of the fire are regulated by the rate at which fuel is added. People in Nepal often use a short bamboo pipe to blow the fire if they want to increase the heat output quickly. Alternatively, the amount of heat absorbed by the pot can be adjusted by raising or lowering it over the fire. This is done by simply shifting the position of the stones.

The open fire can burn almost any size, shape and type of fuel. Where cutting tools are primitive or unavailable to women, it may be the only practical way of using the large pieces of dead wood gathered during fuel collection. This is particularly so in places where much of the available fuel consists of the branches of dried hardwoods which are very difficult to break or cut.

Stones can be positioned to provide a stable support for different shapes

An improved stove may use less wood; but big pieces must be chopped into manageable lengths. An open fire can burn very long pieces of wood, if they are moved into the fire as they are burned.

---

and sizes of pots. Many traditional diets are based on foods which are cooked in the form of heavy paste-like mixtures of flour and water. Among these are tö in Upper Volta and ragi in India, both of which are made from millet flour. These porridges require prolonged stirring in the pot while they are being cooked. This means that the pot must be securely supported and prevented from sliding about; the three stones around the fire can be ideal for this task.

Nevertheless, the open fire can be dirty, dangerous, and unpleasant to use. It makes the dwelling in which it is used difficult to keep clean. As families earn more money, most want to abandon the open fire for some form of stove.

Fuel economy

The three-stone fire is accused of being highly wasteful in energy. There are many reasons for this charge.

When a fire is built between stones openly positioned around it, the heat can radiate laterally out from the burning fuel. Energy is thus lost to the surroundings. The upward flow of heat is only partly obstructed, so much of the hot combustion gases pass freely up around the pot without transferring any of their heat to it.

Any wind greatly increases heat losses. The extra air coming into the fire increases the rate of combustion and hence the heat output. At the same time, the wind blows the hot gases away from the pot. Thus, an unshaded, open fire in a wind uses more fuel but puts less heat into the pot. Even inside, heat losses due to wind can be considerable, particularly if the walls of the dwelling are made of straw or other woven or permeable materials. Even the normal draughts and circulation of air through a dwelling can affect a fire's performance.

On the other hand, much can be done to increase the efficiency of fuel use. Using smaller stones and positioning them closer together — giving a good fit to the pot — can bring dramatic improvements in thermal efficiency. The direct radiation of heat and light from the fire to the surroundings is greatly reduced; so are the effects of wind or draughts. The hot combustion gases transfer more energy to the pot because they pass closely around its sides on their way upwards.

Simple wind screens combat the worst effects of wind. Ghanaian women often fill the spaces between the stones around the fire with mud, while women in the Transkei often cook inside the shelter of low rectangular walls. Also, the fire may be built in the lee of a wall.

Where fuel is scarce, most people take great care with the control of the fire and with the recovery of unburnt fuel when the cooking is finished. In Senegal, "Economy is second nature, and women will quench the fire with water or bury the embers in sand immediately after the cooking is through", according to one survey. Also, even if a cook is using fuel in a thermodynamically inefficient way, not much is being wasted if not much is being used for the job at hand.

However, the thermal efficiency of cooking is only a part of the total question of domestic fuel consumption in the open fire. Once a pot has been lifted, the efficiency of the fire, measured as the transfer of heat to the pot, drops to zero. If the fire is used to provide heat or light at times when cooking is not taking place, then its efficiency can hardly be judged only on the basis of how well it heats pots.

The need for room heat can be an important consideration in places such as Nepal, Peru and Lesotho, where winter temperatures drop below freezing in the higher regions. Even in countries with a generally hot climate, a fire may be needed for keeping warm in the evenings or in the early mornings. In Kenya, there are places where the night temperature drops sharply, and children and old people tend to sit close to the fire in the cooking hut.

Fires can provide domestic lighting, but stoves do not. For many families light is a vital factor in fulfilling deeply felt aspirations. In Kenya, many rural children prepare their school homework by the light of the open fire. The same use of the fire is found in West Bengal, where families are prepared to spend a lot of time or money so that children have light to study by. In Malawi, the open fire is used to supply lighting for 21% of rural families. Very often, the

Mark Edwards/Earthscan

Any breeze robs a pot of heat. Stones and wood have been piled up to shield this fire in Niger from the wind.

---

fire is the only source of light for the woman who is cooking. Thus many families would not trade the open fire for a stove until they had another source of light, be it electricity or kerosene. The provision of kerosene for lighting raises the possibility of kerosene cookers and heaters.

In many African societies the smoke and heat from open fires preserve grains and other foodstuffs stored in the rafters over the fires, while the smoke may keep insects out of the thatched roof and help to preserve the thatch.

Fires play other, less concrete, but no less important roles. In Guatemala, for example, the fire often serves as a social centre for the family, with meals being eaten around the hearth. Women often eat kneeling next to the hearth. In Nepal, the hearth has religious and ritual significance among the higher castes.

In the Third World a fire is more than its thermal efficiency. When a fire is being used as a social or ritual focus, or as a source of heat or light, it is working best when its energy is being radiated as effectively as possible away from the fire towards those who want it. The bigger, hotter, and brighter the blaze, the better it serves its purpose. Improved cooking efficiency may thus come at the expense of equally important attributes of a fire.

Chapter 3

Traditional, "unimproved" stoves

Simple stoves, made to traditional designs, exist in a great many parts of the developing world independent of efforts by either domestic or foreign agencies who want to get "improved" cookstoves into homes. These stoves vary greatly in shape, size and materials of construction; if small local variations are taken into account, there are hundreds of different types.

In many places, it can be hard to decide what is a "stove" and what is an open fire. In Upper Volta, for example, one method of making an open fire in places where stones are hard to find is to cut two intersecting furrows in the ground. The pot is put at the crossing, and the fire is lit beneath it.

There is not a great deal of difference between this arrangement and the traditional Bangladesh *chula*: a short, slightly downward-sloping tunnel dug into the ground, with a pothole cut in the roof at the end. A fire is lit beneath the pot and is kept alight by feeding it with fuel pushed in from the open end of the tunnel. In some cases the fire chamber is made big enough for two potholes. The depth of the fire chamber in these *chulas* is around 40-55 cm (16-22 inches).

Often fires are placed on a waist-high platform and partly surrounded by bricks or mud blocks. These may also have metal bars or a grating spanning the fire as a support for the cooking pot. Although little more than raised open fires, these are generally described as stoves.

The Woodburning Stove Group at Eindhoven University in the Netherlands has provided a formal classification of stoves. In this, stoves, whether traditional or new, are divided into three broad categories: shielded lightweight, shielded heavyweight, and closed heavyweight. Each of these has a further set of sub-categories defined by the number of potholes in the stove. The term "shielded" means that the fire is at least partly enclosed; this is what broadly distinguishes stoves from the open fire and its variations.

The shielded lightweight category covers all portable metal and ceramic stoves. Shielded heavyweight stoves are those constructed of bricks, mud, concrete or other heavy materials, but not equipped with chimneys or draught-control devices. Closed heavyweight stoves are similar to these, but have chimneys, fire doors, dampers or other means of controlling the flow of air through the fire.

Cooking with traditional stoves varies greatly throughout the world, depending on climate, local customs, family structures, the types of food being cooked and the economic status of people. Some meals require long periods of boiling and simmering, others depend on quick heating or frying. Some families may have a supplementary kerosene stove which can be used for making snacks and tea.

In many countries, cooks squat by the stove or sit on a low stool. For many Indian women, this is the only daytime rest they get. In Latin America, cooking is traditionally done standing up. In other areas, the older women wish to continue cooking at ground level, but younger women sometimes prefer to do their work standing.

In hot countries, people frequently take advantage of the portability of a lightweight stove — as they do of the portability of an open fire — and do their cooking in the coolest place available. Among poor Sudanese families, the cooking is usually done outdoors in the courtyard under any shade from the surrounding buildings. As with the open fire, stove cooking may be done outside in the dry season and indoors during the rainy season, or in a special cooking hut.

In fact, where the traditional stove does not offer enough control or flexibility, people may use a fire in addition, or as an alternative.

Open fires may even be used for reasons of fuel economy, as found in a study of several villages in the Andean region of Peru. One of these villages is in a very arid area where there are few trees and great pressure on land resources, with a consequent acute scarcity of fuel. Cooking in this village is done over an open fire on the floor of the kitchen, with the pots carefully positioned to conserve fuel. Yet a traditional local design of stove is well known and used in nearby villages where fuel is more plentiful.

Portable stoves

Light metal stoves — generally used to burn charcoal, though they may also burn wood — are found in most of the urban areas of Africa. They are made by local tinsmiths from scrap metal, often old oil barrels. In Kenya and other East African countries there is the ubiquitous *jiko* — Swahili for "stove" — which appears to have been introduced 50 or 60 years ago.

The *jiko* is normally made in the form of a cylinder about 25 cm (10 in) in diameter and 15 cm (6 in) high, with a perforated metal grate about halfway up inside. Pot supports are fixed to the top edge and project inwards a few centimetres. Fuelling is done by feeding small pieces of charcoal around and under the pot, or by lifting the pot. Ashes are removed through a small side door at the base of the stove. This door can also be used as a means of controlling the flow of air to the grate.

A similar stove made of light sheet steel is used in India, where it is called the *angethie*. It may come with a bucket-like handle, and in some areas is actually made from a bucket and is lined with a layer of mud and cement. This insulates the stove, cutting down on the radiant heat losses from the sides. A cylindrical

24

Forneau malgache stoves, made from sheet metal scrap, come in a variety of shapes and sizes. Ashes are removed through a hole in the base, and this opening can also be used to control the supply of air to the fire.

Aprovecho Institute

stove with an outside wall of sheet steel and an inner lining of cement is also used in Indonesia. The grate consists of iron bars fixed into the cement lining.

Another type of double-skin charcoal stove, often referred to as the *Thai bucket*, is widely used in Thailand. It has an outer metal skin made from a bucket, often complete with the handle for carrying it. The inner lining is made in the same bucket-shape out of fired clay. It has a grate, usually made of fired clay; because the grate often breaks, people buy a number of spare grates with the stove. Ashes are extracted through an opening in the side, below the grate. The space between the inner pottery lining and the outer metal skin is filled with an insulating material such as ash from rice husks. The top edge and the opening for removing ashes are sealed with cement. This stove was developed in Thailand around 1920, and is now used all over the country.

In the Sudan, charcoal stoves are frequently made by local craftsmen from cld paint or kerosene tins. In this type of stove, a grate made of wire mesh or woven metal strips is fitted across the open top of the tin, or set slightly below it. The ashes fall through the grate to the bottom of the tin where they are removed through a hole in the side.

In French-speaking West Africa, the common charcoal stove is called the *fourneau malgache*, *foyer malgache* or *feu malgache* (all of which refer to "Madagascar fires" or "furnaces"). In some of the English-speaking West African countries, it is in fact called the *Malagasi* or *Madagascar* stove. This too is made from sheet metal scrap and is found in a variety of shapes and sizes. Most have a circular or square stem surmounted by a dish-shaped head which holds the fire. The grate is set part way up the stem, or the coals may lie in the lower part of the head itself with the pot resting directly on them, or supported above them on a metal grid. In other cases, the cooking pots have short legs which support them clear of the charcoal. Ashes fall through the grate into the stem and are removed through a hole in the side. This opening can also be used to control the supply of air to the fire. As in the case of the *jiko*, fuel is added by pushing small pieces under the pot, or by lifting it. Because of the dish shape of the head, charcoal automatically slides downwards into the centre of the fire, and for this reason these stoves are sometimes described as "self-stoking".

A very simple metal wood stove is found in parts of Upper Volta. It consists of a metal sheet bent into the form of a squat cylinder which sits around the fire, acting as a wind shield. There is a hole in the side through which fuel can be pushed into the fire. Three short metal lugs project inwards from the top to support the pot.

The single pot ceramic *chula*, (also spelt choola, chullah, or choolah), is found in many Asian countries. *Chula* is the Hindi word for a wood-burning stove. These stoves are usually cylindrical, around 15 cm (6 in) in diameter and 10 cm (4 in) high. They tend to weigh about five kg (11 lb) and are easily portable. Fuel is added through a hole in the side at ground level. When wood is the fuel, quite long sticks can be used. One end of the stick is placed in the fire and the remainder extends outwards along the ground. Straw, dung, and other materials are also burned. Some designs of portable *chulas* are made with a grate; this is always the case when charcoal is the fuel.

The Indonesian version of the *chula* is called the *keren*, and a traditional Japanese model which has a lid is called the *Kamado* cooker. Some designs of ceramic stoves found in India and the Philippines incorporate a base which projects outwards from the stove on the side of the opening through which the fuel is added. This increases the stability of the stove, as well as retaining the ashes and preventing them spilling out onto the floor. It can also accommodate a pot and keep it warm while other food is being cooked.

The Bangladesh ceramic *chula* is simply a fired clay vessel, rather like a cooking pot itself. Upward projections from the top edge provide a raised seating for a pot. The fire is in the bottom of the *chula*, and fuel is pushed in from the top through the space between pot and stove.

A ceramic charcoal stove, similar in shape to the *fourneau malgache*, is widely used in Indonesia. It stands about 17 cm (7 in) high and has a dish-shaped top with a diameter of about 23 cms (9 in) in which the fire sits. The grate is in the centre, at the lowest part of the dish. The stem is about 12 cm (5 in) in diameter and has an opening at ground level for ash-removal. A small, elegant-looking charcoal stove of the same type, of soapstone, is found in Somalia.

Tom Learmonth

This two-pothole traditional mud stove in Nepal burns rice straw. The lids behind the cooking pots cover the potholes to save heat when the holes are not being used.

A simple stove specially made for burning rice husks is used in Bali. It consists of an open-top oil drum with a hole in the side at ground level. Before loading it with rice husks, a thick stick is laid horizontally across the bottom of the drum from the centre outwards through the hole in the side. Another stick is held vertically at the centre. The rice husks are then packed tightly around the two sticks. When a fire is required, the sticks are removed, leaving a passage for air through the opening in the side, across the base, and up through the centre of the fuel bed. The fire is kindled at the bottom, and provides about two hours' burning.

A ceramic form of this rice husk stove is found in Yogyakarta in Central Java. Rectangular or dome-shaped brick versions which rely, in the same way, on a centre hole and a horizontal air inlet formed while packing in the fuel are also found in Bali. Sawdust can also be used as fuel in these stoves.

Portable stoves are commonly lit outdoors. If there is a breeze, this can make the kindling easier, particularly if charcoal is the fuel. Another reason for lighting them outside is to avoid creating smoke inside the dwelling. When the stove is burning properly, and no longer emitting smoke it may be carried inside.

All these portable stoves are made by local tinsmiths or potters and sold locally. Few last very long, usually one or two years; metal deteriorates under the heat from the fire and eventually burns through. Ceramic stoves can be more resistant to heat, but they are extremely vulnerable to damage through being dropped or knocked by the cooking pot.

Fixed stoves

Larger fixed stoves are are usually made of mud, mud and clay bricks, or mud and sand; they are known collectively as mud stoves. Straw, dung, or sometimes wire mesh or pieces of metal may be added as reinforcement. Similar stoves are made of sand and cement.

These stoves are sometimes coated with a thin paste of cowdung or cement to prevent them from cracking. They are usually built on the floor of the dwelling. Some are made directly in their final shape; for others, a solid mud block may be formed, into which the fire chamber, fuel opening and potholes are then cut. Few are fitted with chimneys.

The simplest versions, which provide little more than a partial shielding for the fire, consist of a surround on three sides of the fire with an opening for fuel at the front and room for a single pot. Often two are built side by side, both being lit and used for cooking when the need arises. The pot is sometimes seated on three or more small raised mounds around the pothole. These permit the hot gases and smoke from the fire to escape upwards around the sides of the pot. In other cases the pot fits tightly into the pothole.

Single pot stoves of this type are common in Nepal and in many parts of

India, and are also found in both single and double fire arrangements in Togo. A similar mud stove, consisting of a mud block about 25 cm (10 in) high with openings for two fires, is found in Upper Volta.

Slightly more elaborate versions of this basic design are achieved by putting a cover across the opening for the fuel, providing a complete surround for the bottom of the pot. This cuts the heat loss at the front of the stove and improves the transfer of heat to the pot. Such stoves, built singly or in pairs, are found in some of the rural areas of India.

Purpose-made ceramic linings for the fire chamber may be used in the construction of mud stoves. Old earthenware water vessels are used for the same purpose. After some years, algal and bacterial growth may accumulate in the porous walls of a clay water vessel, making it unusable. To make it into a single pot stove, a hole is broken in its side to provide an opening for fuel, and it is then built into a low mud platform where it acts as firebox and cooking pot support. Several may be placed side by side to allow the use of more than one cooking pot at a time.

Many types of mud stoves have two or more potholes. In some, these are positioned more or less symmetrically over the firechamber, with each pot obtaining roughly the same amount of heat. The potholes may be side by side, as is found in the rural areas in Egypt, Indonesia and other countries. In other types, the potholes are placed one behind the other with the fuel being pushed in from the front. Such stoves are used throughout Indonesia.

In other stoves of this type, only one of the holes is directly over the fire chamber. The other potholes are connected to the fire chamber by a short passage formed in the body of the stove. When a pot is placed on the main pothole, the hot gases from the fire are forced through the flue to the secondary potholes. These may be to the side of the firebox, such as the two and three-pot stoves in Tumkur district near Bangalore, India; or they may be one behind the other, as with the ceramic version found in the urban areas of Indonesia.

In all of these designs, the pot placed over the centre of the fire heats more quickly than those on the secondary potholes. When only one pot is in use for cooking, the other potholes are usually covered to prevent a waste of heat, or they may be used for heating water. In parts of India, a pot of milk is kept warm on the stove throughout the day, as this keeps it from spoiling in the hot climate.

Some fixed stoves, versions of which are found in Nigeria, have a chamber below the firebox for storing and drying fuel. One elaborate combined stove found in mid-Java has a two-pot firebox for charcoal and a somewhat larger two-pot firebox for wood. Storage for wood and charcoal is provided in separate chambers in the base.

Most of these fixed stoves of mud, or sand and clay, are constructed by their owners. They are built in accordance with local designs, materials, and customs. There may be wider cultural or non-technical considerations. A survey of some villages close to the Gulf of Cambay in Gujarat, north-west India, found that the direction in which a *chula* points can be important. Some families are unwilling to build it facing east or west; but whether this is a religious matter or simply the remnants of a traditional adaptation to the direction of the prevailing wind from the coast is not clear.

Mud stoves without ceramic linings do not tend to last very long. The heat of the fire causes them to crack, sometimes quite severely. Pieces break off from the side of the potholes under the weight or impact of the pot, and from around the opening through which the fuel is added. They are also vulnerable to water spilling on them, and can disintegrate if wet thoroughly.

Most mud stoves last one or two years, but many are worthless after only six months. Repairs are difficult because it is hard to bond new material to the old. Towards the end of its working life, a badly deteriorated mud stove performs like a simple open fire.

Some extremely durable homemade stoves, however, are found in Central Java. These are made of limestone or siltstone slabs with a mud covering. They are rectangular and have two potholes, one behind the other, with an opening for fuel in the front. A marketed variety of this stove is found in the same area. This is made of metamorphic siltstone which is quarried and shaped by artisans. The stoves are sold in the marketplace in the form of five interlocking pieces which are assembled in the home. The lifetime of both types is about 10 years.

In Latin America, one of the commonest stoves is called the *poyo campesino*, or *poyo con plancha*. This consists of a clay or brick platform — "poyo" is the Spanish word for platform — on top of which are placed three brick or adobe surrounds for the fire. A metal plate is laid across the top and acts as the cooking surface. This plate may be solid or it may have two or three potholes fitted with removable covers. In other instances a metal grill may be used instead of the plate. This type of stove is common in the Andean villages of Peru.

The stove is often fitted with a chimney or vent above it to remove smoke from the dwelling, and some 135,000 families — about 15% of the families using wood as their cooking fuel — were using such stoves with chimneys in Guatemala in 1979. People interviewed said that they used about the same amount of fuel with the stove as with the open fire. The principal reasons they gave for using a stove were that it was more convenient, and produced less smoke and dirt than an open fire.

One of the most elaborate traditional systems for cooking and domestic central heating is found in Korea and parts of China. Here, the stove and the method of extracting smoke and combustion gases are built as an integral part of the basic structure of the house. Cooking is done on a slab which sits above the hearth; wood and coal are the usual fuels. The exhaust gases from the fire are drawn off through a flue, which passes into an open space below the floor of the dwelling. The sub-floor heat serves to warm the dwelling.

These exhaust gases are then discharged through a chimney, which may be built as a separate free-standing unit slightly away from the house. The chimney is often heavily insulated with straw to prevent excessive cooling of the flue gases during winter. Such cooling would reduce the drawing power of the chimney and cause condensation of the tars and moisture contained in the flue gases.

Fuel economy

It is as hard to determine what constitutes the "thermal efficiency" of a traditional stove as it is to decide what that expression means when applied to an open fire. When the stove is required for room heating as well as cooking, heat which escapes around the pot or through the fuel opening into the dwelling is not being used inefficiently, but is helping to heat the room.

Similarly, when the woman cooking in the dark interior of a dwelling has to rely on the stove for illumination, light radiating through the open front of the stove cannot be described as "wasted". Closing the stove front would save fuel, but another source of lighting would be needed.

But traditional stoves can, nevertheless, be wasteful of energy. Heavy mud or brick stoves lack an effective way to control the flow of air through the fire. Combustion is normally regulated by the way fuel is added. Sometimes a piece of metal, such as a pot lid, may be placed across the opening in the fire chamber in order to reduce the draught. But the air supply in traditional stoves is often above or below the optimum required for the most efficient fuel use.

As regards the transfer of heat to the pot, in stoves with one pothole the amount of heat absorbed by the pot depends greatly on where it is in relation to the upward stream of hot gases from the fire. In some stoves, the pot fits tightly into the pothole, while in others it rests on small mounds in or around the pothole; the second arrangement means that hot gases escape through gaps around the pot. The efficiency, in practice, depends greatly upon how well any particular pot is matched to the configuration of the stove.

Portable metal stoves such as the *jiko* and *fourneau malgache* also waste energy due to the poor insulating properties of the metal itself. A great deal of heat is radiated through the stoves' walls to their surroundings. The superior insulating qualities of double-skin stoves such as the *Thai bucket* make them more efficient than the simple metal types.

The energy efficiency of two or three-hole mud stoves should in principle be higher than that of stoves with only one pothole. The presence of additional potholes enables a higher proportion of the heat contained in the combustion gases to be usefully absorbed. But the amount of heat captured at the secondary holes depends on the general working state of the stove, and particularly on the soundness and fit of the pot at the main pothole. The efficiency of the stove also depends heavily on how well the air supply to the fire is controlled.

Despite their deficiencies, however, most traditional stoves can be controlled and operated with care when the need arises. The following by Uno Windblad, a Swedish consultant, describes the use of a traditional stove in Bhutan:

> "Long pieces of fuelwood can be used, and the intensity of heat can be varied by shifting the wood from one hole to the other, or by pulling out the fuel from the firebox. Bhutanese farmers in this way operate their traditional stoves with considerable skill. Heating is fast, as a large part of the cooking pot is directly exposed to the fire. Slow cooking can be carried out on coals raked into the shallow pit in front of the firebox. These coals also warm the people gathered around the stove in the evening."

Traditional stoves continue in use not because they are of particularly efficient design, but because they meet the demands put upon them by the user. On any one count they may be seriously deficient, and users are generally willing to point out the defects in their stoves to anyone who questions them. But they are used because they fulfil all of their necessary functions adequately. They are reasonably well suited for cooking the different types of food that must be prepared on them. They are appropriate to the climate in which they are used and to the economic conditions of the people who rely on them.

They are an effective compromise between the often conflicting requirements of utility, economy, convenience and compatibility with local housing. Better stoves may be invented for testing in a laboratory, but it is usually more difficult to invent a better stove for everyday cooking in a Third World village.

Chapter 4

Domestic fuels and how they are used

Wood is the main domestic fuel of some two billion of the planet's people - about half the population — and is the dominant fuel in the rural Third World. In the cities, charcoal is often the main cooking fuel; but wood is also used, particularly by the poor.

Wood dominates as a fuel even in areas where it is scarce. In some Asian countries associated with long and severe woodfuel scarcities, wood is still commonly used. In Nepal, where deforestation has reached devastating proportions in recent years, 97% of the domestic energy is supplied by wood. In Upper Volta, where deforestation is also severe, the dependence is over 90%. Though India is well-known for the widespread use of dung, wood is still the most widely used rural cooking fuel. Over large areas it is the only fuel used.

In Guatemala, 80% of the rural population use wood as their sole cooking fuel; a further 15% also rely on it, but supplement it with kerosene or bottled gas. In Malawi, it supplies virtually all the fuel used for cooking, domestic heating and water heating for bathing. The same has been reported in Tanzania, Ethiopia, Somalia and other countries.

The collection of fuel is often the responsibility of women, especially in the sub-Saharan African countries where wood is usually carried home in headloads. Children often help. But in parts of Senegal and Upper Volta men also lend a hand bringing home wood. But often the men collect wood mainly for sale.

"The behaviour of a man who uses a motor bike, a cart or even an automobile for this (wood collection) is significantly different, since for him bringing home wood is no longer considered a sign of inferiority and submisson to his wife. On the contrary, it is the recognition of a new situation, that is that wood has become a rare and therefore valuable commodity deserving the attention of the male sex!", according to a 1981 article by Jacqueline Ki-Zerbo.

Men are responsible for providing the family fuel supplies in some of the Latin American and Asian countries; in these the wood is often transported in animal-drawn carts. In Nicaragua, it may be carried on the back of a horse. Men also help in Zimbabwe, where a bullock-drawn sledge called a "scotch cart" is sometimes used for bringing the wood home.

The loads of firewood which women carry on their heads can be extremely

The typical Zimbabwe bullock-drawn sledge, called a "scotch cart", is being used to carry wood back home.

---

heavy. An average of 20 kilograms (44 lbs) seems common in many African countries. Loads of up to 33 kg (73 lbs) are carried in Tanzania and one of 39.5 kg (87 lbs) was noted in KwaZulu.

In most places, dried twigs and branches which have died a natural death are collected. In places where axes or machetes are not available, people must collect what they can pick up or break off by hand. Even where cutting tools are available, people rarely fell whole trees for domestic firewood, but they may cut live branches and twigs. These must be stored and dried. This drying, which must also be done when wood is collected during the rainy season, can be a serious problem where houses are small and storage space is limited.

The availability of wood does not preclude the use of other fuels. Coconut shells and maize (corn) cobs, which burn in roughly the same way as wood, are freely used when they are available. Straw, twigs, leaves and other light combustible materials are used for kindling; they may also be used for cooking quick snacks or making tea, as they quickly provide a hot flame. Millet straw, which women in Upper Volta frequently dampen before use in order to slow down the rate at which it burns, is another common fuel of this type.

Dung

Dung is rarely used in areas where wood is readily available to all families. Its use is likely to be an indication that there are wood supply problems, for some families at least.

In India, some 73 million tonnes of dung are used annually, accounting for perhaps a third of the total consumption of traditional biomass fuels in the country. Much dung is used in Bangladesh, where it accounts for about a quarter of traditional fuel use. Dung is burned in many other countries, but generally on a smaller and less widespread scale.

In India, agricultural residues are generally added to the dung to act as a binder so that the cakes into which it is made do not fall apart. On some farms, extra animal fodder is strewn on the stable floor and trampled into the dung by the livestock. Or it may be mixed in with the dung by hand when the dung cakes are being made.

Investigations into cooking patterns in villages in Lesotho and the South African "homelands" of KwaZulu and the Transkei show shifts between wood and dung as fuels, and how their use varies throughout the year. Lesotho, because of its high altitude, has cold winters; it also has very few trees. Conditions in the Transkei, at a lower altitude, are similar, but winters are milder. In both, for fuelwood people rely on small bushes, which are gathered whole, including the leaves, roots and stems. They are principally burned during the summer, but are also used in the spring and autumn.

In winter, dung is the main fuel. Some is collected in the fields by women and young girls, while the rest is dug twice yearly from the kraals close to the dwellings in which the milk cattle are kept overnight. This dung is dried and made into bricks, each weighing about 3.4 kg (7.5 lbs). When it is used as fuel, it is kindled with brushwood and burned in a perforated tin. Although it gives off a thick, acrid smoke, it is preferred to the brushwood as a winter fuel because it burns slowly and provides warmth for the dwelling as well as heat for cooking. The bushes, in contrast, burn quickly without creating coals or giving off much heat. Another reason for the use of dung is the difficulty of collecting the bushes in winter.

In KwaZulu villages, good quality fuelwood is freely obtainable nearby and is the sole fuel for cooking. Women walk several kilometres to carry large pieces home in headloads. Kindling is collected nearer the dwellings. Although dung is as readily available as in Lesotho and the Transkei, it is not used for fuel at all.

In the southern Peruvian altiplano near the town of Nuñoa, dung has been the traditional fuel for centuries, and is used with considerable skill. Here, at an elevation of 4,000m (13,100 ft), the climate is cold and dry; plants grow slowly and small, and there are very few trees. People live by subsistence farming, and herding sheep, cattle and llamas. Dung is used for fuel by 87% of the population.

Dung cakes are made and stacked to dry under the Indian sun. Use of dung as fuel usually indicates that wood is not readily available in the area.

Cattle dung is collected from the fields and corrals near the houses. People know how to get the best from it. It is kindled with grass, lights readily, and burns with a hot flame and little smoke. Llama dung is also readily obtainable around the homes, and its collection is easier because these animals deposit their dung in communal piles. This dung produces a hot and even fire, but one which is difficult to maintain, so it is used along with cattle dung. Sheep dung, which produces an acrid smoke, is not used for fuel but is the basic source of fertilizer.

The average family uses about 30 kg (66 lbs) of dung daily — equivalent to the production of 19 cattle or 75 llamas. The dung produces almost as much heat as the locally available wood.

Commercial wood markets and the use of kerosene

When fuelwood is bought and sold as a commercial good, it tends to be viewed differently from when it is gathered without any payment. The transition from the use of non-commercial to commercial fuel can have a considerable impact on domestic energy consumption patterns. Where no commercial fuelwood market exists and people pick up their fuel for free, they will tend to respond to a scarcity of fuelwood by shifting to other non-commercial fuels: dung or straw, for instance. They will be reluctant to invest money in improving fuel supplies such as by planting trees, or in reducing consumption through the use of a more efficient stove.

In contrast, people who normally buy their wood have a cash incentive to invest in fuel-saving devices when wood becomes scarce and more expensive. They also have a motive to plant trees. They may be willing to switch to other purchased fuels, either as supplements for wood or as complete substitutes.

The coexistence of non-commercial and commercial fuelwood supplies in the same village or community is common. In the Bhavangar district of Gujarat, India, the poorest people use fuelwood which they gather from waste lands. It is usually obtained from the babul tree *(Prosopis cineraria)*, the wood of which is hard and thorny and thus difficult to collect and prepare for the fire. Only the poor are willing to go to this trouble. People who have jobs in the nearby towns, or those who are working as teachers in the villages tend to use part of their income to escape from the necessity of collecting their own fuel. They buy firewood which has been neatly cut; they also tend to use kerosene stoves.

This use of a kerosene stove as an additional means of cooking is found in many countries. Around 40% of the Fijian households which rely on the open fire supplement it on occasions with a kerosene stove, and there is a close relationship between kerosene consumption for cooking and household

A kerosene stove is useful when firewood is scarce, and is handy for preparing quick meals and tea. As their incomes rise, people become less dependent on woodfuel. This woman is using the kerosene stove to supplement her single pot chula (note the dung cake on the left).

income. When income increases by 1%, there is a 1.5% increase in kerosene consumption.

However, cooking with kerosene can cause severe problems for some people. For example, the traditional tortillas of Latin America are difficult to cook on a kerosene stove since they need a wide and uniformly heated cooking plate. Balancing a pot on top of a light kerosene stove while stirring a thick porridge or stew can be difficult and dangerous. Kerosene can give an unpleasant smell or taste to foods, such as the Indian chapattis (flat bread).

Kerosene stoves can easily flare up unexpectedly if handled carelessly, and make the wearing of saris made of highly flammable synthetic fabrics dangerous. Kerosene stoves can also be extremely dangerous if they are knocked over and their fuel spills.

Despite its disadvantages, kerosene can improve the living conditions of those who switch to it. Being subsequently without it may be felt as a serious deprivation. In West Bengal, India, the position for rural dwellers when kerosene runs short is described by A. Ramesh: "In the rural areas, instead of coal and kerosene, wood, hay, fallen leaves and dry plants are used as fuel. But the nights are the worst, as no kerosene means dark nights". During a kerosene shortage, people may burn more wood in the open fire to provide light for the family in the evenings.

A shift to kerosene saves the labour involved in collecting wood and cuts down on the amount of smoke in the dwelling. Kerosene stoves are particularly useful for meals which do not require much preparation, such as breakfast, or for making tea. In the Transkei, the younger women tend to prefer kerosene, which they regard as a modern fuel, whereas older women still accept the collection and use of fuelwood as their traditional duty. Surveys in Indonesia show that people rank kerosene second to wood in their preferences as a general cooking fuel, but they also regard it as a valuable supplementary fuel for cooking small meals.

The move to kerosene is part of a general shift away from wood as incomes increase — a shift clearly evident in most developing countries. The more money people make, the more they can spend on kerosene, bottled gas, charcoal, electricity, coal, or other commercial fuels. The final form of this shift — in which virtually no fuelwood is used — is found in most industrial countries.

This changing pattern of fuel use will have a big effect on any attempt to introduce new cooking stoves. The economic conditions under which they can be most easily introduced may also be those in which people are already considering changing away from wood. For instance, in India:

> "It is a moot point whether the poorer strata will ever consider *wood-*burning stoves (even costlier and improved ones!) as an elevation in their standard of life when they are aware of the fact that the more affluent sections of their society always prefer cleaner and more convenient cooking fuels to wood. The poor know that there is a hierarchy of cooking fuels and they view changes from fuelwood to charcoal to kerosene to electricity or gas as steps in the improvement of the quality of their life", according to Indian rural development expert A.K. Reddy.

The switch from non-commercial woodfuel to kerosene and other commercial fuels is so complex that its effect on stove promotion is difficult to evaluate without a great deal of information about the context in which it occurs. The shift towards kerosene has frequently been seen as evidence of hardship and scarcity of woodfuel. In fact, it may equally be evidence of an increase in prosperity and a desire to escape from the use of wood. Once the transition has been made, for whatever reason, people are likely to be reluctant to reverse it. But they do switch back to wood and other fuels when kerosene becomes expensive or unattainable.

Quantities of fuel used

It is impossible to get reliable and detailed information on domestic energy consumption without careful, long-term monitoring. Ideally, this requires living in the community and observing people unobtrusively, so that their normal patterns of fuel use are not disturbed. Seasonal differences in climate and fuel consumption must also be established if a full picture is to be obtained. Such surveys are expensive.

Few of the studies carried out to date provide information on year-round consumption. Most refer to dry-season consumption, for the simple reason that many communities are almost totally inaccessible during the rainy season. Thus the significance of any seasonal variations in fuel consumption cannot be established in many of the areas surveyed.

Few surveys have a detailed breakdown of fuel consumption by "end uses". How much fuel is used exclusively for cooking? For heating? For lighting, or in keeping the fire burning for convenience or for social reasons?

It is often difficult or impossible to compare the findings in the available published data. Methods of measuring and recording the fuel consumption of fires and stoves under field conditions vary between researchers. Some reports give the quantities by volumes, others by weight. Sometimes they are given in bundles, headloads, cartloads, or even in local vernacular measures. Fuel consumption may be specified per head, or per family, with no indication of family size. The quantities may be those used per day, week, year; or the amount required to cook a meal. It is rarely stated whether the fuel is air dry, oven dry or simply as it has been gathered.

The American researcher Deanna Donovan has analysed 50 estimates of domestic fuel consumption in Nepal. There is such a wide variability in the average sizes of families assumed, in the moisture content and density taken for wood, in the weights assigned to traditional volume measures and various other factors affecting the estimates of per capita consumption that it is almost impossible to find a valid basis for comparison between the studies. From the highest to the lowest, the estimates of fuel consumption vary by a factor of no less than 67.

Thus much of what is said and written about the quantities of traditional fuels consumed is based upon impressions and ambiguous evidence, rather than firmly established data. Nevertheless, a sufficient amount of information now exists for a more detailed picture of fuel supply and consumption patterns to be drawn than was possible even a few years ago. It provides a background, albeit somewhat hazy at times, against which the general fuel-saving prospects for stoves can begin to be examined.

There are also wide variations in energy consumption even within the same community. This has nothing to do with incompatibility of data; it is just that under similar circumstances different families consume very different amounts of fuel. This phenomenon is not restricted to developing countries.

Energy surveys in industrial countries also show large differences in consumption between apparently identical households.

In Fiji the climate is generally warm and most families do not use the fire as a means of heating their dwellings. Cooking is done over an open fire on the ground, or on a raised platform. Lighting and a small amount of supplementary cooking are provided by kerosene. Fuelwood is plentiful and generally obtained free. The results of a survey of villages and isolated farms showed an average annual consumption of 353 kg (778 lbs) per head taken over the whole survey area. Breaking the figures down to the level of the separate villages, the average varied from a low of 283 kg (624 lbs) per head up to 413 kg (910 lbs). The only area with significant use of fires for domestic heating was, in fact, that with the lowest total wood consumption: 283 kg (624 lbs) per head. In this village, heating is used for an average of five months in the year.

In Upper Volta, a similar variability in the quantities of fuel used in the open fire has been found. In April and May 1981, some field tests were carried out to begin calculating the fuel savings obtainable from introducing improved stoves. The total fuel consumption of 17 families living in squatter dwellings on the outskirts of Ouagadougou was monitored for a week. Wood was the only fuel used.

The average daily consumption turned out to be 0.93 kg (2 lbs) per head, with a range of 0.63-1.36 kg (1.38-3 lbs), equivalent to 230-496 kg (507-1,093 lbs) annually. A follow-up survey two weeks later showed per capita consumption to have fallen to 0.84 kg (1.85 lbs) per day, with a range of 0.49-1.21 kg (1.08-2.67 lbs), or 179-441 kg (395-972 lbs) per year. Between the highest and the lowest per capita consumption over the full survey period there was therefore an almost threefold variation across the 17 apparently similar families.

A large survey in Nicaragua has shown that consumption levels vary widely between different areas. The survey covered 518 small farm families and was carried out in 13 different departments of the country. Cooking was done almost entirely with fuelwood. Annual consumption ranged from 1,100 to 2,865 kg (2,425-6,316 lbs) per person. The majority of people used some kind of stove, though in some departments up to 33% of people cooked on the open fire. No definite relationship between the average consumption and the percentage of people using the open fire was obtained. However, in the two areas of highest per capita consumption the use of stoves was virtually universal.

In Guatemala, cooking is done on the open fire or on a simple raised stove. Lighting is provided by kerosene, propane or electricity. Virtually the whole country, apart from a small high-altitude zone, has a tropical or sub-tropical climate. The average total consumption of fuelwood in the rural areas, on an oven-dry basis, has been estimated to be 1,650 kg (3,637 lbs) per head each year. This was based on a sampling of the amounts purchased by people who used wood as their only cooking fuel. In families where wood and kerosene were used, the total annual wood consumption was 1,250 kg (2,755 lbs) per head, together with an average of 4.4 gallons (20 litres) of kerosene.

Surveys in southern African villages have demonstrated the relationship between fuel consumption and the availability of woodfuel. In Lesotho and the Transkei, the annual per capita fuelwood consumption was 288 and 271 kg (635, 597 lbs), respectively. In Lesotho this was supplemented by 260 kg (573 lbs) of dung per year, and in the Transkei by 80 kg (176 lbs). The calorific value of the dung proved to be about two-thirds that of the wood, so the total equivalent fuelwood consumption was about 440 and 315 kg (970, 694 lbs) per year. The proportion of the total fuel used for heating, in both cases, was estimated to be about 30%.

In contrast with these villages where wood is scarce, much larger amounts of fuel were used in the village surveyed in KwaZulu. Here wood is abundant, and annual consumption was found to be 1,124 kg (2,478 lbs) per head. This is despite the fact that the climate is considerably milder than Lesotho and the Transkei, though about 13% of fuel goes for heating. In all three cases, the proportion of total energy consumption used for lighting, which was supplied by kerosene, was only 1-2%.

The village of Kwemzitu in the mountainous region of north-eastern Tanzania offers another example of high wood consumption when it is freely available. The village of about 200 inhabitants is beside a forest reserve from which the people are allowed to take wood, most of it gathered from a place one hour's walk from the village.

The mornings and evenings are cool. The morning fire is lit at first light and used to prepare a breakfast meal of maize porridge. The fire may be allowed to burn until the last person leaves for work around 8 a.m. Another fire is lit at noon and may be left burning through the afternoon if the weather is cold. The evening fire is lit about 5 p.m. and may be kept burning until 11 p.m. The annual wood consumption per head was found to be 1,636 kg (3,607 lbs) per year.

On the Indonesian island of Lombok, where fuelwood supplies are limited to those brought by small traders from the mountain areas, the average annual consumption was equivalent to 327 kg (721 lbs) per head, of which 19% was kerosene. Some 40% of the people used bark, leaves, and other biomass material as well as wood. In Klaten, in Central Java, supplies of wood are also limited and all fuel must be bought or gathered from people's own land. Here the consumption was 447 kg (985 lbs) per head, with 24% being in the form of kerosene. Around 40% of the people also used bark, leaves, and biomass residues.

Fuelwood supplies are abundant in the area of Luwu in South Sulawesi, but this abundance causes a difference in the pattern of fuel use rather than the total quantity consumed. Here, the amount of wood equivalent consumed averaged 489 kg (1,078 lbs) per head per week. The amount of kerosene used was just 2% of the total and the remainder was in the form of fuelwood proper with no use of biomass residues.

The amount of fuel used for space heating is extremely difficult to estimate. In Nepal, where the climate in the higher regions is very cold in winter, up to 35% of the domestic fuel supply for a family may be used for heating, depending on the weather and the availability of fuel. When wood is readily available, huge quantities are used for heating, but as supplies become scarce, domestic heating is restricted to little more than the period during which cooking is taking place. Under these circumstances, there is obviously a substantial suppressed demand for fuel for heating.

The amount of fuel used for space and water heating may also have to do with levels of income. Some studies of rural domestic energy consumption in three zones of peasant agriculture in Mexico show that the proportion of total fuel used for cooking falls as income rises. In the lowest income bracket, around 90% of the total was for cooking, with only 3% each used for water and space heating. In the higher income bracket, 34-42% of the energy was used for water heating, and 9-15% was used for space heating. Surveys in Zimbabwe have shown that the proportion of fuel used for heating can range from 19% to 34% of the total fuel consumed.

Fuel is often used for purposes other than meeting the family's direct cooking and heating needs, and such uses can play havoc with domestic energy surveys. Such uses include boiling animal food, baking bread or cooking sweets for sale, pottery making and small-scale brick burning. Religious and family celebrations can consume substantial amounts of fuel. The domestic brewing of beer is also a common activity across the whole developing world. And even warm countries, such as Upper Volta, may use a lot of hot water for bathing.

Some of these activities can be done on an improved stove; some cannot. But they must be considered in any study trying to decide whether better stoves will save firewood. As David French wrote of his findings in Malawi: "Among the uses of wood in Malawi which would partially or wholly *not* be done on improved stoves are beer brewing, heating water in large quantities for bathing, smoking fish or meat, curing tobacco, cooking potatoes or pumpkins in extra-large pots, space heating, lighting. It is important to specify these things, since they sharply limit the potential for saving wood by introducing improved stoves".

Relationships between consumption and scarcity

One of the apparent paradoxes of the fuelwood problem is that in many areas where forests are being rapidly cut, fuelwood itself is often plentiful. Thus a great deal of wood may be used in the mountain and high forest areas of Africa, Asia and Latin America. In arid areas, where biomass productivity is naturally low, less fuel is used. In some of the Asian countries where population densities are high and wood is difficult to obtain, consumption may be lower still.

This means that where fuel is becoming scarce, those feeling the pinch will already be trying to save fuel. Try though they may, such people will not be able to save much fuel by adopting an improved stove. Those who will save more by using a better stove are those who are using more wood — that is, those who live where wood is plentiful.

But where fuelwood is plentiful, people have less incentive to use new stoves, especially as fuelwood is generally easier to burn over a three-stone fire than in a stove which may require small pieces of wood.

Chapter 5

Designing "new" stoves

The designer of a new stove does not have a free hand. If a stove is to have any chance of being widely adopted it must be compatible with the technical skills and economic status of those who will make and use it. People must be able to cook their usual foods on it at least as conveniently as they do by traditional methods.

In many cases, designers may want to base their new stove on the types of stoves already used in the area. Improved versions of these will not have to face the barrier of complete unfamiliarity. In other cases, a radically different approach may be more effective. People in developing countries are not necessarily averse to fundamental change in their cooking methods, as the widespread adoption of kerosene and bottled gas proves. If the move from an open fire or traditional stove to a particular kind of new stove offers clear advantages, people may well be willing to make the change.

So a knowledge of what local people want in a stove is crucial for a designer. However, some requirements of a practical stove design will be in conflict with others. Low cost may be achieved at the expense of durability; energy efficiency may require complex manufacture and operation. Stove design is not an exact science, but a craft requiring compromise, judgement and sensitivity to the needs and abilities of the intended cooks.

The key difficulty is achieving a balance among efficiency, flexibility in use, cost, durability, and simplicity. The different designs reflect the different judgements made on the right balance among these considerations for different areas and groups of consumers.

Combustion and the transfer of heat

A well-designed stove must allow fuel to burn completely. For this to happen, the fire chamber must be of the correct size for the fuel being used, and there must be an adequate supply of air.

Wood and vegetable residues can burn well on the ground or on the base of a stove. So stoves which use these fuels are usually made without grates. But tests have shown that a grate can improve combustion and lead to a higher energy efficiency. Some woodstove designs therefore incorporate one, though this does increase their complexity and expense. Charcoal stoves need a grate. Bulky fuels such as straw, leaves and twigs — even though they may be used only part of the year — require a larger fire chamber than if the only consideration were conserving the smaller fuel normally used.

Designers disagree on the importance of what is called "secondary air". This is air admitted above the heart of the fire into the flame zone, as opposed to the "primary air" drawn into the base of the fire. If there is any shortage of air in this zone during combustion, the secondary air can compensate for it and improve combustion efficiency. Secondary air inlets in the form of small holes are incorporated in some ceramic stoves, and are found in some extremely old traditional designs.

The efficiency of the combustion in the fire chamber is only the starting point in the process of effectively transferring energy to the pot. It is the transfer of heat to the cooking pot which gives designers their major problems. The two most important mechanisms by which this happens are radiation and convection. Their relative contributions will vary depending on the cooking pot, where it is relative to the fire and the behaviour of the fire itself.

Energy radiates from the embers and flames of a fire in all directions: through the rest of the fuelbed, onto the surrounding stones or stove walls and onto the bottom of the cooking pot. When the radiation encounters an object, such as the outside of the cooking pot, it is absorbed or reflected in differing degrees depending on the colour, texture and other characteristics of the surface. Surfaces which have been heated in this way also reradiate heat in all directions.

Convective heat transfer is the name used for the various processes by which the hot combustion gases give a proportion of their energy to the pot as they flow past it. As their temperature rises in the fire, the density of these gases is reduced and they become more buoyant. They flow upwards around the sides and bottom of the cooking pot. The rising gases are then replaced by cooler air flowing inwards towards the fire, thus causing the natural draught with which the fire replenishes its oxygen supply.

The amount of convective heat transfer depends on such factors as the upward speed of the gases, the shape of the cooking pot and how the gas flow extends around it. Hence, where the pot is in relation to the fire can make a substantial difference to its heat absorption.

Improving the open fire

It is often claimed that the open fire has an "efficiency" in the range 3-8%, meaning that only 3-8% of the energy in the fuel is heating the pot and cooking the food. But without further information, such statements are almost meaningless. In indicating that the open fire is necessarily wasteful and inefficient, they can also be seriously misleading.

Laboratory experiments conducted at Princeton University, USA, found very much higher efficiencies for the open fire than normally quoted. Using wood of moisture content up to 25%, and setting the pot 11 cm (4.3 in) above the base of the fire, resulted in an average efficiency around 19%. Another set

46

The low bank around this open fire can save wood. Hot combustion gases transfer more energy to this beer-making pot in Upper Volta because they pass closely around its sides on their way upwards.

---

of experiments resulted in similarly high results, and researchers found a close relationship between efficiency and the height of the pot above the fire. In these tests, oven-dry wood was used. The maximum efficiency of 23.2% was found at a height of 7.5 cm (2.9 in), and this fell to 11% when the height was 22 cm (8.7 in).

Tests by the Woodburning Stove Group at Eindhoven University investigated the effects of moisture content and the height of the pan above the fire. The results were similar to those at Princeton. The group obtained efficiencies of over 20% under a variety of conditions, and even with wood of 25% moisture content got an efficiency of 18.8%.

Experiments on shielded fires equipped with grates and burning oven-dry wood obtained even higher efficiencies — between 23.4% and 33.8%.

These were laboratory tests under controlled conditions. There were no random draughts. Care was taken both with running and monitoring the experiments and with the feeding of fuel and control of the fire. The results do not imply that such efficiencies could be reproduced on a village cookstove.

A "real" open fire is vulnerable to breezes and draughts. So the simplest and most obvious improvement is an enclosure around it, with an opening at the front through which fuel can be fed. This protects the fire from the wind and cuts down on the radiant losses. Some of the heat absorbed by the shielding is radiated back into the fire and onto the pot instead of being lost to the environment. Although a fire enclosed in this way is frequently described as a stove, it is really nothing more than a shielded fire.

Shielding a fire also directs the flow of combustion gases more closely around the pot, increasing the amount of heat absorbed by it. Normally the pot is put on top of the shielding, but the improvement in heat transfer can be even greater if the pot is set lower over the fire so that it is partly or wholly enclosed by the shield. Obviously, the pot must fit the shield, and this need restricts the size of pots which can be used.

In principle, the open fire can achieve the same efficiency as a shielded fire or a simple one-pot stove if it is built in a place sheltered from the wind and is carefully tended and refuelled. But the shielded fire, a simple and natural evolutionary step from the open fire, works better and reduces the dirt and mess from the fire. It also offers some protection for children. It seems to be used, in one traditional form or another, virtually everywhere.

Improved metal stoves

Light metal stoves, used mainly for burning charcoal, are found everywhere in the urban areas of Africa, and to some extent also in Asia and Latin America. Most are well suited to their function and reasonably efficient. In fact, metal and ceramic charcoal stoves are better suited to their fuel than are most fuelwood stoves. Fireclay and ceramic models are comparatively efficient.

But light metal stoves lose heat through their walls. So designers have sought ways of insulating them to cut down on these losses. Fuel efficiency aside, insulated stoves are less likely to seriously burn those who touch them.

The Intermediate Technology Development Group has also tested the traditional East African *jiko* in comparison with a number of experimental insulated stoves. The experimental stoves were all made with double skins. Two types, which were like the *Thai bucket* in that they had pottery liners and an ash insulating layer, gave efficiencies of 25-30%. The efficiency found for the *jiko* was 20%. In some other tests conducted in Tanzania a *Thai bucket* was found to use just over half the fuel of a *jiko*. The *UMEME* stove, which has an insulating layer of soil and is being developed in Nairobi by UNICEF, has shown a test efficiency of 34-36%. But it is designed so that the pot fits down into it, as opposed to resting on top as in the normal *jiko* and *Thai bucket* types. So it is limited to a particular size and shape of pot.

The *Thai bucket* and other such double skinned stoves have proved the technology of the insulated charcoal stove. Their popularity in several countries shows that their practical merits are recognised. The problems in introducing them to other countries are related to the lack of technical skills available for making them; the established patterns of stove manufacture and the willingness of craftsmen to consider change; their economics; and the degree to which householders can be persuaded that the energy savings are worth the extra cost of switching from a traditional stove.

Stoves with two or more potholes

The normal *single-pot* stove in which the pot sits on top, rather than being sunk into the pothole, has a major limitation. Once the gases have escaped past the pot, there is no practical way to capture their remaining energy for cooking purposes.

Thus most improved stove designs are based on the use of two or more potholes. By channelling the hot gases past a second, third, or even fourth pothole, the opportunities for capturing heat which would otherwise escape to the atmosphere are greatly enhanced. When cooking requires only one pothole, the others can be covered or used for heating water or keeping food warm.

Designers have tested different configurations of potholes and gas flows in order to maximise the energy performance without reducing the convenience and practicality of the stove. They want to extract as much heat as possible from the gases before they leave the stove, without making the normal cooking tasks slower or more difficult. By altering the size of the gap around potholes, for example, the proportion of gases directed towards other potholes can be adjusted.

Cooks, as well as designers, are concerned about getting the correct balance in the amounts of heat delivered to the various potholes. Reducing the heat supplied to the main pothole may increase the overall energy efficiency of the stove, but cooking will take longer. The dimensions and positions of the fire-chamber, gas passages, baffles and other parts of the stove are crucial in the design of multiple pothole stoves. The efficiency and general performance of a badly designed stove, or one which has deteriorated severely, may be worse than those of a reasonably tended open or shielded fire.

Choice of materials

Designers must make hard choices between cost and long-term performance. Stoves made of mud, or clay and sand, can be made by their owners from free, locally obtainable materials. Examples include the *Lorena* stove designed for Guatemala and the *Ban ak Suuf* stoves designed for Senegal. These are both made from a mixture of about 75% sand and 25% clay. But they tend not to last very long. Six months hard use without repairs can leave a *Ban ak Suuf* stove looking almost unrecognisable. Also, the dimensions of owner-built stoves vary more than those made by trained artisans, and efficiency suffers.

Gerald Foley/Earthscan

These two Ban ak Suuf stoves have deteriorated badly. Wear around the left pothole and its fuel opening was caused by everyday use. After only 18 months, half of the stove below has crumbled away completely.

Gerald Foley/Earthscan

Another free material from which stoves can be built is called "banco". Similar to the adobe found in the Middle East, Latin America and elsewhere, banco — mainly clay with some sand added — is a familiar local building material in Niger. Animal manure and some straw or grass are also added to provide fibrous reinforcement. Water is added to this mixture, which is left for a few days before being used to build a stove.

Banco is harder and more durable than the *Lorena* mixture. But it is more difficult to prepare, and the stove takes a long time to dry out. Traditional local expertise, such as is available in Niger, cannot be called upon in many other areas.

Once people are ready to pay for stoves, designers can make them more robust. Cement can be added to the building material or applied to the stove surface to increase its strength and durability. Metal or ceramic reinforcements can be used to strengthen critical points such as the fire chamber or the edges of potholes. *Lorena* stoves with a cement coating on the outside are used in Guatemala. Durability may also be increased by making stoves of brick, such as the *Nouna* stove in Upper Volta.

Different geographical areas may require stoves of different masses. Stoves with a high mass absorb a lot of heat just after they have been lit. So at first they are relatively inefficient in cooking because the heat is going into the stove and not the pot. Gradually the stove reaches a state of equilibrium with the fire, and its efficiency increases. But heavy stoves will always be slower and less economical than light stoves when used for short periods of cooking. When habitually used for long periods, heavy stoves respond quickly when fuel is added and thus perform as rapidly and economically as light stoves.

Because of its high mass and insulating qualities, the temperature of the outside walls of a heavy stove such as the *Lorena* rarely goes above about 40° centigrade (104° F). It thus heats a kitchen less than an open fire. In hot climates this is an advantage; but in colder areas, comfort may depend on the heat radiated from the stove.

At the end of the day, when cooking is finished, a heavy stove will contain a lot of accumulated heat. If the fire is banked or extinguished, heat will be gradually emitted during the evening and night. This will be an advantage if the climate is cold at night, and the stove is in a place where people can sit or sleep close to it.

In general, high mass stoves are best used in places where the fire is kept alight most of the time, or where room heating is needed. They are less suitable where cooking is intermittent and it is not customary to keep the fire alight for long periods.

Other stove designs have been proposed in which a plate formed from sheet iron, iron plate, reinforced concrete or ceramic material might be used to form the top of the stove and support the pots. These may be accurately made and are durable, but in most places they would cost more than people are willing to pay.

The use of ceramic inserts or liners for stoves is showing promise in Sri Lanka, Indonesia and Nepal. The dimensions of the fire chamber, internal flues and potholes are critical in determining the energy efficiency and general performance of stoves. Often these are incorrectly formed by badly trained or insufficiently skilled builders. Also, they deteriorate and change with wear. The use of a ceramic insert can reduce or even eliminate these problems.

Trained potters can make the inserts accurately, and they can be distributed or sold to householders or organisations promoting stoves. This reduces the task of stove building to the simple task of embedding the ceramic inserts in mud. The use of the insert cuts down on the workload of stove programme field workers. Hence more stoves can be built for the same investment in extension services.

Draught control

Control of the draught through a stove is essential if the fire is to burn properly and the gases are to move past the various potholes. This is often achieved by a hinged or sliding door in front of the firebox. It is kept fully open while the fire is being lit or when the maximum draught is wanted. It is then adjusted as required to control the air flow during cooking. With careful use of such a door, a fire can be precisely controlled.

But in practice these doors, which are either hinged or fit into channels along which they slide, often become detached or damaged with wear. Even in dwellings where lamps are available, cooking is often done by the light from the fire. A closed fire door can mean no light for cooking; so those using the stove may leave the door open when energy efficiency requires otherwise.

If fire doors are used, wood must be cut into short pieces. This is harder work than collecting the extra wood which would be needed by a less efficient fire. So the use of fire doors is often abandoned after a time, and the stove becomes less efficient.

The air flow also may be controlled by an adjustable damper, perhaps in the form of a metal plate set into the lower part of the chimney itself, or placed so that it can block the gas passageway through the body of the stove. Such dampers often fall into disrepair, and it is difficult to teach people how to use them properly.

Smoke and pollution in the kitchen

There has been little research on what constitutes an acceptable standard for air quality in a kitchen. Most countries do have legal limits on the amounts of suspended dust particles permissible in a working environment, and pollutants from wood fire such as carbon monoxide, benzo(a)pyrene (BaP),

Good ventilation is needed to free the house of smoke. This household in Senegal has another solution: it is building this Louga stove in a separate hut.

and nitrogen dioxide are often subject to individual limits. But such limits are rarely enforced in Third World kitchens.

The limited number of air quality measurements taken in typical developing country kitchens show that the concentration of pollutants is often high by most recognised global standards: "Indeed, they show that cooks receive a larger total dose than residents of the dirtiest urban environments, and receive a much higher dose than is implied by the World Health Organization's recommended level, or any national public standards", according to a 1983 report by K.R. Smith of the Resource Systems Institute, Honolulu. "The women cooks are inhaling as much BaP as if they smoked 20 packs of cigarettes per day."

Few medical studies have been carried out on the actual effects of smoke on people cooking on open fires or traditional stoves. Nevertheless, many health problems are associated with spending time in a smoky environment. Bronchitis and other respiratory diseases can be caused by smoke, and are certainly aggravated by it. Smoke can damage the eyes. Some of the complex hydrocarbon molecules found in smoke may also cause cancer.

Smoke is also a nuisance. It and the soot associated with it soil clothes, furniture, wall and ceiling coverings and other household possessions. It can coat electric light bulbs and disrupt rural electrification programmes. Smoke is actually hampering the introduction of rural electric lighting in Papua New Guinea because of the way it fouls the lamp fittings.

The following description by Uno Winblad of the situation in Bhutan could probably be applied in many other places:

> "The most obvious disadvantage of the traditional stove is the smoke. Deposits of smoke and tar rapidly blacken the interior of the kitchen and partly also the other rooms of the house, as well as clothes and household belongings. Smoke irritates the eyes and respiratory system. In order to ventilate away smoke and carbon dioxide it is necessary to provide good ventilation in the house, thus letting in cold air. An alternative sometimes used in Bhutan is to place the kitchen in a separate building. The main building can then be kept free from smoke, but it is an extremely expensive solution. The smoke still irritates those who work in the kitchen".

Chimneys

Though chimneys can minimise smoke, putting them on stoves can bring problems. Fitting a chimney to a stove changes the flow of air through it. The nature of these changes depends on the shape of the stove, the dimensions of the chimney, the position of the outlet, and the weather conditions, particularly the direction and strength of the wind. But generally, chimneys increase the air flow through the fire, so it burns better and is easier to light. But unless the air flow through the fire can be controlled, the fire may burn more fiercely than required, thus wasting fuel.

Badly fitted, wrongly placed or incorrectly shaped chimneys can reduce the flow of air, making the fire difficult to light. There may be trouble in keeping the fire going, and smoke may blow back into the dwelling. Considerable trial and error may be needed to get the chimney right.

Chimneys can increase the fire hazard by overheating or discharging sparks. Metal chimneys are particularly prone to overheating.

Chimneys made of brick or clay can be quite heavy. For example, the three-metre (10 ft) mud-sand chimney used with some improved stoves in Malawi weighs 45 kg (100 lbs). It is hard to fix and secure such a chimney so that it will not fall in high winds, especially if the dwelling itself is light and flimsy. In earthquake-prone areas such as Central America, heavy chimneys can be extremely dangerous. The reconstruction work in Guatemala following the 1976 earthquake made a special point of avoiding them.

Chimneys can increase stove costs and put them out of the reach of many people. A ceramic or metal chimney may cost as much as, or more than, the rest of the stove. A thin metal chimney may last only six months; a ceramic one is likely to last two years or more. People often do not bother to replace a chimney once it has worn out.

If a chimney is not cleaned regularly, it may become blocked, and soot and tars can build up into a fire hazard. Anyone introducing stoves with chimneys will need to train people to clean them.

A chimney can thus be a mixed blessing, and installing one so that it performs satisfactorily throughout the different seasons requires both skill and detailed local experience. The question of whether or not a chimney should be used can only be resolved in the context of local conditions. The fact that few developing country dwellings have chimneys may show the real problems that can arise in their installation and use.

An alternative is short chimneys discharging indoors at ceiling or eaves level. These are cheaper and are less likely to cause problems of excessive draught. Where smoke is required to cure meat or to protect the roof against termites, the short chimney may be appropriate. By discharging the smoke above head level it harnesses any benefits of the smoke while avoiding the worst of its discomforts.

Measuring efficiency

One way a designer can test the "efficiency" of his design is to measure the thermal "efficiency" of the new stove. In the strict scientific sense, efficiency is a measure of the proportion of the total energy which is usefully employed in a thermodynamic system. In the case of a cooking stove or fire, it is the proportion of the original energy in the fuel which is used in heating the pot and cooking the food.

The main difficulty in measuring the efficiency of fires and stoves is the large number of factors which can distort the conclusions of any series of tests. If test results are to be used as a basis on which design or general programme decisions are made, they must be truly comparable with each other. Every significant item being investigated needs to be specified and measured exactly. Such testing methods can require expensive and elaborate laboratory equipment. One group specialising in such detailed laboratory measurements is the Woodburning Stove Group at Eindhoven University in the Netherlands. Their work has done much to establish the precise significance of the different factors which influence the performance of stoves and fires.

The moisture content of the fuel used in a test is one important variable, as the energy yield of newly cut wood may be less than half that of dried wood. Other wood variables affecting energy release include species, density, size and shape of pieces used and the rate at which they are fed into the fire.

A fire changes over time. Just after a fire has been lit, much of the energy released goes to heat the surrounding stones or stove, as well as to dry and raise the temperature of the remainder of the fuel. The energy thus consumed will be high in relation to that absorbed in the cooking, and the efficiency with which cooking is done will appear low. After this initial phase, the cooking pot begins to absorb a higher proportion of the energy released from the fuel, and the cooking efficiency appears to increase.

A clay pot is a much better insulator of the food inside it than a metal one. Hence, during a given time, the food inside a clay pot will absorb less energy from the fire than would a metal one. Comparability between tests requires that the same type of pots are used.

If a pot has no lid, a great deal of the heat absorbed by the food is rapidly lost again. In some stove and open fire tests in Fiji it was found that when the lids were left off, the water did not come to the boil at all; but it did boil when the lids were replaced. Any evaporation from the pot must be measured. It takes eight times as much energy to boil a given quantity of water away as it does to bring it to the boil initially. So neglecting the energy absorbed in vaporising water during a test can distort the results.

The same method of lighting the fire must be used for all tests, as this can affect the manner in which the combustion takes hold and the heat is transferred to the pot in the initial stages. Testers must also make the appropriate allowances for the energy remaining in the unburnt wood or charcoal left over after the test.

Perhaps the most important variable of all is the person carrying out the tests. Other things being as far as possible equal, it has been found that different testers, or stove users, can produce quite different test results. Testers must also be careful to carry out the various tasks in exactly the same way in consecutive tests. Improper loading of the fuel and inconsistent attention to the fire can lead to a difference of up to 100% between tests.

None of these problems is insurmountable under laboratory conditions. The difficulties arise when tests are carried out in the field, when it is difficult or impossible to measure and record all the significant variables.

In actual cooking, the rate at which the heat is supplied is as important as the quantity of heat supplied. When a dish such as a stew is being cooked, the cook wants to bring it to a boil relatively quickly. After that, only enough heat to keep it gently simmering is needed. The fire which efficiently delivers heat but takes an interminable time to bring a large pot to the boil may be thermally efficient, but is neither efficient nor practical from the viewpoint of the cook. The fire which continues to supply more heat than is needed after the boiling point has been reached is clearly wasting fuel, no matter how thermodynamically efficient it is in doing so.

Laboratory tests can provide comparisons between different stoves or methods of using them, and can help researchers examine the effects of altering any variable. They are design tools rather than predictors of field performance. The laboratory measurement of efficiencies can never be used to predict how much fuel stoves are likely to save in rural villages.

Testing performance

Given the limited applicability of laboratory efficiency tests, stove researchers need a practical method of evaluating stoves and fires under working conditions. The results of many past stove tests are unreliable because testers did not take adequate account of the variables discussed above.

In December 1982, 13 stove experts from 10 countries met at the offices of Volunteers in Technical Assistance (VITA) in Arlington, Virginia, US. An agreement was reached on provisional international standards for carrying out and reporting three basic tests. Detailed instructions on the exact method of carrying out each test, and standard forms for reporting the results are provided in a publication resulting from the meeting, "Testing the Efficiency of Wood Burning Cookstoves: Provisional International Standards".

The Water Boiling Test (WBT) is intended to provide a quick comparison of the performance of different stoves, or the same stove under different operating conditions. It measures the amount of fuel and the time taken to bring a specified quantity of water to the boil and is carried out in two phases. In the first, the water is heated as quickly as possible from the ambient temperature up to boiling, and kept at the same high level of power output from the fire for 15 more minutes. This is then followed by a low power phase in which the output from the fire is reduced to the lowest level needed to keep the water within 2° centigrade of boiling for another hour. In effect, this is a simulated cooking test.

The Controlled Cooking Test (CCT) is meant to compare the fuel used and the time spent in cooking a real meal on different stoves, and to determine whether a stove can cook the range of meals normally prepared in the area in which it is to be introduced. This test is usually conducted in a laboratory or field demonstration centre. A number of typical local meals are cooked separately, and the time taken and the quantity of fuel used in each case are recorded. The test is carried out at least five times for each type of meal, and then repeated on a different stove or on the traditional stove or fire used in the area.

The Kitchen Performance Test (KPT) measures the relative quantities of fuelwood consumed by two stoves when they are used under normal household conditions. It is a prolonged test, which needs the cooperation of the family concerned, and is normally undertaken only after the previous two tests have been completed. It requires a minimum of at least five participating households, and is meant to measure the quantity of fuel consumed by a family using the stove as part of its normal domestic routine over a period of about a week. The role of the tester is to record details of the family and the meals cooked, and to measure the amounts of wood used every day.

As this test establishes whether a new stove can save fuel in practice, it should be done in the early stages of any stove programme. But it is a difficult test both to carry out and interpret.

But even with these carefully designed testing procedures, there are pitfalls. For example, the introduction of a new stove into a household may change kitchen routines. The new behaviour may be temporary, and people may become more wasteful once the test is over. Or the family may choose to cook more on the new stove. So improvements in efficiency, if they occur, may not be revealed as a reduction in fuel consumption. Or the stove may impose additional costs on the family which need to be set against any benefits which the stove brings. For example, the enclosure of the fire in a stove may reduce the amount of light available, so that a kerosene lamp must be used; or extra time may have to be spent cutting wood into small pieces to fit the stove.

In a warning about such problems, the testing guidelines state:

> "It may be tempting to use the results of the KPT to estimate the fuel saving potential of the new stove before it is widely accepted and used. For this purpose, however, the test would have to be greatly expanded to include:
> * many more households, carefully selected to be representative of the regional population;
> * a period of time that includes all major seasons;
> * a study of stove deterioration rates and repair records;
> * an economic analysis demonstrating the economic attractiveness of the stove to both the user and producer".

A performance test is not a fuel consumption survey. Performance tests provide a method of comparing different types of stoves, and can be used as a basis for deciding on changes in design or stove operating procedures. They do not provide information on the actual quantities of fuel that would be consumed if a large number of people bought and used the stove.

Conclusions

It is possible to design stoves which have a higher efficiency than those in normal traditional use. But it is harder to translate these into practical models which people will use. Much depends upon the economic level and technical capabilities in the area.

Also, few poor people still able to collect fuel without payment will be willing to invest money in the energy-saving potential of a stove. In such places, simple, home-made, very low cost stoves will be the most appropriate. Their main disadvantage is their low durability.

Where fuel is commercially traded, it may be easier to introduce more costly — and more energy-efficient — stoves. But the type of stove and the skills required in its manufacture and use must be closely matched to the people building and using them.

Thus the design of such a stove must be based upon what people themselves want, rather than what outsiders believe they should want. This requires open-minded enquiry among the people who are intended to use the stove, and a genuine deferment to their wishes. US anthropologist Marilyn Hoskins wrote in 1979:

"One frequently hears questions like, 'If you had a stove that used only half the fuel you currently use, would you cook on it?' This is not the real question. The question might more accurately be, 'If you had a stove that used less fuel would you be willing to chop your wood into 20-cm (8-inch) lengths, control the damper and clean the flue?' Technicians are not in a position to evaluate these types of trade-offs. Only if the woman is given the necessary information will she be able to make an accurate evaluation of the new stove's potential for being adopted".

The process of designing and marketing a stove is similar to that which successful manufacturers of most consumer goods engage in before they launch a product. A well-designed stove can provide more than just an opportunity for saving fuel. It can be used to control or avoid the emission of smoke; it can aid in the heating or cooling of the cooking area. It can be an object which people are pleased to have in their homes. But if a stove does not please the customer, it will not be bought or used.

Chapter 6

Stove programmes: past and present

One of the earliest improved stoves, the *Magan chula*, was developed in 1947 at the All-India Village Industries Association, Maganwadi. It was made of clay, with straw or dung added as a binder. The stove was formed from a solid stove-shaped block; when it dried, knives and trowels were used to carve the firebox, three potholes and the flue, and then a ceramic chimney was fitted. In many respects, this stove was a direct precursor of the *Lorena* mud stove of Guatemala.

New design features included curved rather than straight ducts, which were intended to promote turbulent gas flow and a more effective heat transfer to the pots; a metal grate; and a combined ash pit and additional air vent under the grate. The stove could burn wood, charcoal or cowdung.

During the 1950s and 1960s, further work was carried out in India on improved versions of traditional mud *chulas*, the work part of a general rural development programme inspired by Gandhi after the country's independence. The new *chulas* were intended to save fuel and to alleviate eye and lung complaints by decreasing smoke in the cooking area. A 1964 survey by the National Building Organisation of India showed that, in the period 1950-64, over 55 improved designs had been developed in the country. Many were very similar, and only a few were widely publicised.

Dr S.P. Raju made a major effort in the 1950s to disseminate improved stoves on a massive scale. Working at the Hyderabad Engineering Research Laboratory (HERL), he developed new stoves made of mud, or brick and mud plastered with fine earth. These were basically an L-shaped duct with one, two or three potholes. At the end of the duct was a wider and deeper opening to hold a large pot for heating water. The combustion gases and smoke went out through a chimney of ceramic pipe, bricks or metal. These *HERL chulas* established the pattern for many later stove designs throughout the world.

The *chulas* were introduced and publicised through Raju's 1953 pamphlet *Smokeless Kitchens for the Millions*, which was often revised and reprinted. One of Raju's main objectives was fuel-savings. He may well have been the first to suggest that more efficient stoves could combat deforestation, and to provide a numerical justification for his arguments. He said improved *chulas* could save about 20 million trees per year in India, and 80 million worldwide.

But he had other advantages in mind. In a note to his pamphlet's "women readers" he wrote:

"You are working for the emancipation of women. Do not forget the millions of your sisters in the bondage of criminally unhygienic kitchens. Do not rest till you have fought and won for every housewife in India her Five Freedoms of the Kitchen:
1. Freedom from Smoke
2. Freedom from Soot
3. Freedom from Heat
4. Freedom from Waste
5. Freedom from Fire Risk."

In 1957, the pottery section at Gandhiniketan Ashram, near Maduraia, developed a "portable" *Magan chula*. This was very different from the original *Magan chula*, as it was made from ceramic parts made by local potters. The stove consisted of three ceramic cylinders lined up and linked by ceramic pipes at mid-height. The third cylinder was connected to a chimney. The stove was assembled by fitting the various parts together and plastering the joints firmly. It was generally embedded in mud to provide insulation and protection for the pottery.

Pots rested on top of each cylinder, and were heated by the fire which was lit on a metal grate in the cylinder furthest from the chimney. There were no dampers to control the air flow, but clay baffles under the second and third pots helped to increase their heat absorption. It could burn charcoal, or wood of any length if the sticks were allowed to extend out along the ground from the grate.

This stove has become established in the area around the Gandhiniketan Ashram, where the Ashram potters produce 100-200 of them monthly. In 1980, 2,100 such stoves were sold each month in nearby Madurai by the S.E.V. Trading Company.

In Indonesia, Hans Singer spent three months in 1959 working at the Regional Housing Centre, Bandung, as a consultant for the Swiss government and the UN Food and Agriculture Organization (FAO). His mandate was to improve the cooking devices in use. His *Report to the Government of Indonesia on Improvement of Fuelwood Cooking Stoves and Economy in Fuelwood Consumption* (1961) was the result of this work.

First, he conducted tests on the calorific contents of local woods and charcoal. The tests led him to emphasise the importance of using only thoroughly dried wood for cooking. He estimated that the use of dry rather than green firewood would result in country-wide fuel savings of about 10%, or 10 million cubic metres (353 million cubic feet) of wood per year. He also ran efficiency tests on traditional Indonesian woodburning and charcoal stoves. The woodstoves all had open hearths and were made of clay, or bricks and clay. He found all to be inefficient in their energy consumption, but judged six charcoal stoves much more efficient.

He designed a series of improved stoves, now known as *Singer Stoves*.

Mark Edwards/Earthscan

A woven shield is being constructed around this three-stone fire in Haiti to protect it from wind and help to contain the smoke.

Model I had three potholes arranged along a bent flue. Model II had three potholes lined up along a straight flue. Model III was for smaller families and had only two potholes. All had chimneys equipped with special caps to keep out rainwater, and all proved more efficient than traditional stoves.

Singer's stoves were similar to Raju's *chulas*, but their metal damper was in the entrance to the firebox. The pots were sunk 3-5 cm (1.2-2 in) into the gas stream for best operation. All three models could be built in either a low (30 cm; 12 in) or a high (70 cm; 28 in) version. The high stoves had a compartment for drying firewood.

Singer worked to improve the normal one-year lifespan of traditional stoves. Trials showed that burnt or sun-dried bricks lasted longer than clay, which was not only short-lived but was suitable only for the low stoves.

Singer suggested new stove designs for the mass catering required in institutions such as hospitals or service camps. Before his departure at the end of his contract, he conducted three two-day courses in stove construction and gave cooking demonstrations. He expected local volunteers to continue promoting his designs, but this did not happen. Although the Singer stove designs were widely published, there is no record of them having been adopted in Indonesia or elsewhere.

Another early stove was the *Smokeless Ghanaian Oven* built in several villages by the Department of Social Welfare and Community Development in Ghana during the early 1960s. Local clay or laterite was the main building material, with metal rods to strengthen the stove top. The stove was a four-hole *chula*-type built over a wood-storage compartment. To one side was a baking oven, separately fuelled but sharing the same chimney.

This design was publicised by the Canadian Hunger Foundation, the Brace Research Institute, FAO and others. But no more was heard of them until 1977, when the villages in which they had been installed were revisited as part of a rural energy study by Luann Martin. She found that all had fallen into disuse. Villagers said the stoves used more wood than an open fire, and that they were unsuited to local ways of cooking. One woman said that the stove used larger pieces of wood than the open fire, and that she had to walk further to find such wood.

Other complaints were that the potholes were either too small to fit some pots, or were too large and allowed smoke to seep past. Some said the stove was too high to permit the local banku porridge to be stirred conveniently. Many women preferred to cook sitting on a small stool next to the fire, rather than standing. Some women left the unused potholes uncovered during cooking, with the result that smoke and heat escaped. Local women had apparently not been consulted in the stove's design.

Another Indian stove design of the 1960s was the *PRAI Improved Chula*, which was developed and introduced by the Planning and Resource Action Institute, Lucknow, India, in 1969. The *PRAI* stove has two potholes and a chimney. Sometimes a metal damper is fitted behind the second pot position.

This stove is reported to have been widely adopted by middle-class families in Kerala state. A three-year programme to introduce it into villages in Lucknow was begun in 1976, but only 25 stoves were installed. In 1980, half of these were still in use.

Current programmes

Recent years have seen the development and promotion of many new stoves in different countries. About 100 programmes are now under way or are in the planning or discussion stages. But most of these are very recent, and many have not systematically evaluated their work or published results. Most of the available information is fragmentary, incomplete and even contradictory. The following provides an outline description of what is happening, rather than a systematic and detailed evaluation.

Guatemala

One of the earliest and best known stove programmes followed the 1976 earthquakes in the Guatemalan highlands. Demand for wood as a building

Ariane van Buren

Constructing a Lorena stove under a CEMAT training programme in Guatemala. A mixture of clay and sand – both widely available locally – is laid on a brick base. The completed stove is often decorated to add to the appearance of the kitchen.

Aprovecho Institute

material for reconstruction in this heavily populated area, already suffering badly from deforestation and erosion, created a severe shortage of woodfuel.

The *Lorena* stove design and building technique were developed by Ianto Evans and Donald Wharton of the Aprovecho Institute, working with the Estacion Experimental Choqui, Quezaltenango. (The Oregon-based Aprovecho Institute concentrates on small-scale, ecologically sound technologies and development initiatives.) The design was derived from the Ghanaian and *HERL chulas*, but was made from a clay and sand mixture, which are widely available local materials. This mixture, about three parts sand and one part clay, gave the stove its name; a combination of the Spanish words "lodo" (clay) and "arena" (sand).

The typical *Lorena* stove had three potholes, a metal damper, a chimney, and a built-in container for heating water with waste heat. The stove stood about 76-91 cm (30-36 in) high, and was around one metre (1 yard) square. Builders used layer upon layer of the *Lorena* material to form the solid block from which the potholes and internal flue were later carved.

The designers wanted to develop a stove that conserved wood, controlled smoke and provided protection for the cook and her children. The *Lorena* design was said to be able to use a wider range of fuels than the open hearth, since it provided a more controlled combustion of low quality fuels in the oxygen-poor atmosphere of the Guatemalan Highlands. Developers claimed that the stove saved 50-75%, and possibly more, of the fuelwood normally used for cooking.

The *Lorena* stove evoked enormous early enthusiasm. It was described as "sophisticated in concept, yet easy to build and affordable by all". It was hoped that the stove would spark a resurgence in local craftsmanship. People would be taught about the materials and the method of construction in two-and-a-half day training sessions and would then build stoves for their families and neighbours.

The main promoters were private, church and government organisations in rural areas. By 1980, the Centre of Middle American Studies and Appropriate Technology (CEMAT) estimated that 1,800 stoves had been built under its training programmes. To date, some 6,000 *Lorena* stoves have been built in Guatemala under the various programmes being run by public and private institutions.

But the stoves have their disadvantages. Their large mass makes them inflexible and sluggish to warm up. In some houses, open fires are lit alongside the stove on cold mornings. The clay and sand mixture often crumbles badly. The metal chimneys fitted to the stoves are expensive, at times tripling the stove's price.

Inadequate extension work meant some stoves were badly built. The early training programmes did not explain the underlying principles of construction and operation. Many of the participants in the training sessions were men who proved unable to teach their wives how to use the new stove.

In spite of such difficulties, the Guatemalan *Lorena* programme has been judged a considerable success. The primary advantage cited by families interviewed on their attitudes to their new stoves was the reduction of smoke. Wood savings, the ability to cook somewhere besides the floor, and the constant availability of hot water were also noted as advantages. The fact that the stove, which is often highly decorated and carefully looked after, adds to the appearance of the kitchen was also frequently mentioned in surveys.

Senegal

The *Ban ak Suuf* programme in Senegal is based on the *Lorena* stove-building techniques. This national programme got off to a rapid start in April 1980, when consultants from the Aprovecho Institute visited Senegal to develop a suitable stove design for the country. Several large *Lorena* stoves of Guatemalan design proved unsuitable; but, with the help of local people, smaller models of a similar sand and clay mix were made. These were called *Ban ak Suuf* stoves after the local Walof language words for clay and sand.

Following this visit, a national programme was launched by the Centre for Study and Research on Renewable Energy (CERER) at the University of Dakar, with funding from the US Agency for International Development (USAID). Local men and women are taught by training teams to construct the stoves. These trainees are expected to be able to build additional stoves for neighbours or friends. Later, a field agent visits new stove installations to instruct participants in their use.

Karen Nelson, a US Peace Corps volunteer, worked on the project from 1980 to 1982 and helped spread information and coordinate the work on a national basis. Many other Peace Corps volunteers in Senegal during that time were also engaged in the programme.

The most popular stove is a small chimneyless model known as the *Louga* or women's stove. It consists of a thick circular shield enclosing a single pot resting on three supports, with a front opening to feed the wood into the fire. It is simply a substantial shielded fire.

Louga stoves are designed and built mainly by women. Coumba Gaye is a prominent local worker who has trained many women in the method of making them, and the stove is often called after her. Baobab leaves are often boiled to produce a resin which is applied to the stove to protect it against rain. Other *Ban ak Suuf* stoves with two potholes and chimneys, or designed for use with charcoal, are also being built.

But the *Ban ak Suuf* material is especially susceptible to rain or wetting. Within a year or so of installation, many stoves have almost completely crumbled away.

In 1982, members of CERER and Peace Corps volunteers evaluated the programme. Some 5,000 stoves had been built, about a third of which had chimneys. It was found that 65% of the total were in regular use, although 20% of these were judged to be in a bad condition. The best results were obtained with the *Louga* stoves, of which 77% were still in regular use.

The *Ban ak Suuf* programme has generated considerable interest in Senegal. It has also enhanced local skills and confidence among women. The following observations are by Cheryl Fattibene, who has worked in the programme at a local level for the past three years:

> "The positive aspects of participation in a group activity such as stove building can have a tremendous impact on village women. Most women are amazed that men's participation in the actual "main d'oeuvre" isn't obligatory and that they can do it on their own. Even if *Ban ak Suuf* stoves have somewhat of a short life expectancy, all the preliminary work, organisation, and participation that leads to the actual construction is an accomplishment worth noting in terms of community development. Women can realise a goal in a new domain and the skills they acquire in doing it can then be used in the future."

Upper Volta

Upper Volta has been the scene of many stove programmes. Ouagadougou is now the headquarters of the Permanent Interstate Committee for Drought Control in the Sahel (CILSS), which is a regional stove coordination programme; and the government has created an Inter-Ministerial Technical Commission charged with the study, construction and diffusion of improved stoves. The Service "Foyers Améliorés" (Improved Stoves Service) coordinates the different programmes concerned with fuelwood use, and is the working body of the Technical Commission for Improved Stoves created by the "Direction de l'Aménagement Forestier et du Reboisement" (Directorate of Forest Management and Reforestation).

The West German Forestry Mission stove programme operates in Ouagadougou and Nouna. One model being promoted, the *Nouna* stove, was designed by Rose Marie Kempers while she was living in Chad, and is based on her memories of European World War Two vintage stoves. She worked with local masons to develop a two-pothole stove with a long firebox and a chimney. The stoves seemed popular, and she continued to build them when she moved to Nouna in 1975.

Now, trained masons construct the stoves according to householders' wishes. Three teams of masons build the stoves full-time, while about 25 others have been trained and do the work as a sideline. After the stove has been built, an extension worker calls to supervise the first lighting and advise on maintenance and use.

The most popular version is made of fired bricks and cement. It has two

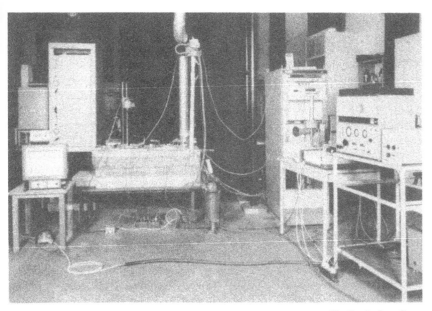

Woodburning Stove Group

A Nouna stove of Upper Volta built specially for testing by the Woodburning Stove Group at Eindhoven University in the Netherlands.

potholes and lasts about two years. When tested at the Netherlands Organisation for Applied Scientific Research (TNO) as part of the work programme of the Woodburning Stove Group at Eindhoven University, Netherlands, it operated at an efficiency of between 15% and 23%. In addition to the two-pot version, there are several models for special purposes: a one-hole institutional model, a high-standing model, and one with three potholes and two fireboxes. The use of cement makes the stove more durable, but the cement must be imported.

Nouna stoves are well-liked, and women find them generally easy to use, attractive in appearance and suitable for cooking local foods. Cooks are particularly pleased that the making of tö, a millet porridge requiring vigorous stirring, is simplified because the pots are held firmly in the potholes. The low height of the stoves allows them to sit on stools while cooking. The stoves are sold at the heavily subsidised price of $20, and buyers tend to be from the middle or upper classes. Most are attracted by the idea of venting the smoke from the kitchen, but have little incentive to save fuel.

The stoves are publicised by special stove demonstration and exhibition centres in larger towns. Demonstrations are also held at fairs, schools, missions, hospitals, community leaders' homes and other prominent places. Over 1,600 stoves had been built by 1982. Problems with the *Nouna* stove include cracking, improperly cured cement and leakage of smoke. Some of the early versions were also extremely wasteful of fuel.

The *Kaya* stove (named after the town where it originated) was developed by Peace Corps volunteer Jonathan Hooper between 1978 and 1980 as a means of stimulating local commercial initiative. "Fourneaux Modernes de Kaya" (FMK — "Modern Stoves of Kaya") is a private enterprise which produces two-hole concrete and block stoves fitted with chimneys. After assembly in a workshop, they are transported to customers' houses. A sand and clay model with concrete rendering is also available, as well as special stoves designed to brew dolo, the local millet beer. The Dutch organisation "Bois de Village" ("Village Woodlots") has given FMK funds to demonstrate several stove models, including the *dolo* stove, at their centres in and around Kaya.

The International Rural Development Fund (AIDR) and the European Development Fund (FED) build stoves in Ouagadougou and Bobo Dioulasso. Stoves with three potholes are constructed by masons using cement and a clay and sand mixture. Efforts are being made to replace the cement with local materials such as banco. AIDR sets up small, autonomous workshops for groups of artisans. The base for the stove and the clay chimney bricks are made in the workshops, and the rest of the stove is built where it will be used. Larger stoves are built for schools, factories and restaurants.

Suspended temporarily for lack of funds early in 1981, the AIDR programme was later funded by the European Economic Community and Interpares (a Canadian non-governmental organisation). Over 18 months ending in 1982, it built 1,200 stoves, most of these placed in urban areas.

Despite the many programmes, only about 5,345 improved stoves had been installed in Upper Volta by the end of 1982.

Kenya

There are many improved stove programmes, in both cities and the countryside, in Kenya. The Kenya Ministry of Energy, the Intermediate Technology Development Group, the West German Society for Technical Cooperation (GTZ), and UNICEF are all involved in this effort. The Kenyan Clay Stoves Working Group at the Kenyatta University has been assisting with testing. The Kenya Energy Non-Governmental Organisations Association (KENGO) has been involved in promotion and has run training workshops and monitored stoves in use. Waclaw Micuta, of the Bellerive Foundation (funded by the Aga Khan), has introduced a number of stove types, two of which, the *Pogbi* and the *Nomad*, are on sale.

A Kenyan version of the *Thai bucket* has been developed by the Ministry of Energy in collaboration with private pottery firms, one of which is a brick and tile company. The stove uses a pottery liner inserted into a metal cladding. The

Traditional and improved jikos on sale in Nairobi, Kenya. The insulating layer between the pottery liner and the outer wall of the improved jikos on the right has been omitted in these versions made by local entrepreneurs.

---

insulated firebox, sturdy metal pot rests, and tight-fitting door make it well-suited to Kenyan cooking. Production of an initial batch of 500 of these stoves was completed in November 1982, and manufacturing is now going on in several places. Following successful initial trials with these stoves, 5,000 metal shells have been commissioned from local artisans by two ceramics companies who are being assisted to set up production units. The companies will fit the liners and then sell the stoves.

Development work is also being carried out on the use of cement and vermiculite to line the traditional charcoal *jiko*, the East African metal cylinder stove. A project for the introduction of 50,000 stoves fitted with vermiculite lining is also under way. There are plans to saturate a selected area of Nairobi with stoves sold at slightly reduced prices and to monitor the overall consumption of charcoal in the area.

Problems arise as more stoves are sold and manufacturing becomes more dispersed. Some local stove-makers and businessmen are cheating, and selling stoves from which the insulating seal between the lining and the metal cladding has been omitted for high prices. But such problems can be seen as proof that market forces are taking over in the dissemination process, which is becoming self-sustaining and passing beyond the control and supervision of the original promoters. While it is too early for final judgements, it appears likely that commercially-produced, insulated metal stoves are on the way to becoming standard household items in much of urban Kenya.

UNICEF has assisted in the making of some 1,000 two-pot mud stoves in rural areas. These have chimneys and are called *Karai* stoves. However, many were so badly made as to be almost useless. Some 200-300 units of a version of a two-pot mud stove have also been built. In addition, an insulated metal stove called the *UMEME* stove (which means "lightning" in Swahili), has been developed in a number of versions. This is designed so that the pot sits tightly in the pothole for maximum energy efficiency.

Indonesia

Yayasan Dian Desa (YDD), a non-governmental organisation based near Yogyakarta, began work on stoves in 1978. First designs were based on the *Lorena* stove, but high mass stoves proved unsuitable for local use. The early programme objectives were to preserve energy and prevent deforestation. These have now been widened to include the reduction of smoke, increasing the cleanliness and appearance of the kitchen and cutting down on cooking time.

During the three years following the early trials with the *Lorena*, a number of stove designs were developed, and about 700 were built in Central Java and Yogyakarta. A smaller three-pot mud stove with a chimney, called the *Katesan*, was developed. Some 5,000 to 6,000 of these were built in one area following some enthusiastic promotion efforts by local government officials. However, most were broken down and unused after three months.

A smaller two-pot *Lorena*-type stove with a chimney, the *Mein Chong*, and a simpler mud stove without a chimney, the *Tungku Lowon*, have subsequently been developed, along with a two-pot ceramic stove, based on an ITDG design called the *Sae* stove. Much work is now being done on ceramic stove liners. Most of the problems with previous models had been caused by incorrect dimensions of critical stove areas. By prefabricating the liner, the problems of quality control are greatly simplified.

Work in Indonesia has shown that local organisations can disseminate stoves on a large scale. The current approach has been described by Marcus Kaufman:

> "A stove has at last been developed which is easily reproducible and highly fuel efficient; the stove has been thoroughly tested prior to any dissemination taking place; an efficient utilisation of Government and private resources has been planned which will enlarge the scope and impact of the programme beyond the limited capabilities of YDD itself; and finally, an integrated educational programme has been designed which will address the needs of field workers, artisans, and villagers in a systematic manner."

The promoters believe that following a few years of sometimes misdirected enthusiasm, the programme is now on a sound footing and ready to make a major impact.

Sri Lanka

The Sarvodaya Shramadana Movement has promoted rural development based on traditional cultural values since 1958. It began to study the promotion of improved stoves in 1979, both as a way to save fuel and to reduce the health hazards from smoke and to improve the hygiene and appearance of the kitchen.

The programme began with the introduction of the *Lorena* stove, but this required maintenance beyond the ability of most owners. Some also used more fuel than the traditional stoves they replaced. Only a small number were built.

The next stove tried, in 1979, was the *Tungku Lowon*. Around 300 of these two-pot, chimneyless mud stoves were built initially, though there was a shortage of trained personnel available for instructing people in stove building. Stoves were also built inaccurately, and deteriorated so quickly that they became inefficient in six to nine months.

Just as in Indonesia, the promoters turned their attention to the design of pottery liners for the *Lowon* stove. After extensive training, the potters were able to produce a sufficiently high quality and accurate product to provide a basis for a dissemination programme. Some 500 liners were installed. After six months, about 90% were still in use, and users claimed savings of cooking time and fuel. The liners are sold for $1.50, and are now being ordered by householders at the rate of about 200 per month.

The Ceylon Institute of Science and Industrial Research (CISIR) Stove Project was initiated in 1978 with the development of new ceramic charcoal stoves for urban use. It wanted to introduce urban dwellers to the use of charcoal, now widely available because of the clearance of land for agriculture and the Mahaweli Dam project. Test marketing of these began in 1980 in Colombo. By mid-1983, 20,000 of these stoves had been produced, and some 4,000 families are using the stoves. Work continues on the development of a double-walled charcoal stove, and on a prefabricated ceramic woodburning stove similar to the Indian portable *Magan chula*.

Nepal

The Research Centre for Applied Science and Technology (RECAST) of Kirtipur, Kathmandu, began its stove programme in September 1981 when a contract was signed between it, the Government, FAO and the UN

Development Programme. The design and testing of stoves was carried out in collaboration with ITDG. Several different ceramic stoves have been developed. One is a double-walled model with a chimney and an air space between the walls to provide insulation. The whole stove is stronger than other ceramic designs, but more difficult to build, heavier and more expensive.

The Centre has also designed pottery liners for mud stoves. In these stoves two pots, the first of which acts as the fire chamber, are connected by a pottery pipe; the second is connected to the chimney. The whole forms a skeleton which is covered with clay, in the same way as the portable *Magan chula*, and is normally referred to as the *insert* type of stove.

The group developed another simple pottery stove with two potholes but without a chimney, called the *Terri stove*. Another single-pot stove with a chimney is being tested. All of these stoves are made by potters in the Kathmandu valley, where there is a long tradition of pottery.

In the beginning of the programme, 90 stoves were distributed to villages representing the four major ethnic groups in central Nepal to determine initial user reaction. Later, a further 690 stoves of the *insert* type were installed. People reported they were pleased with their stoves, which were given to them free. Promoters believe people will purchase a replacement when it becomes necessary in two or three years time. A further 15,000 stoves are to be distributed over 1983-85.

Niger

The Church World Service became interested in stoves at the end of 1979. During 1980, it started a stove project, and in early 1981 several *Lorena*-type stoves were built using banco as the construction material. The *Kaya*-type stove, which is made of sand and cement and fitted with a chimney, was also built. This latter model proved overwhelmingly more popular. When offered the choice between cement and banco, 99% of the families chose the cement type.

A demonstration programme was initiated in Niamey in 1981, sponsored by the Association of Niger Women. During the year, about 380 stoves were installed in the houses of about 20 selected families in each of the 20 different wards in the city. The stove, and the overhead shelter built with it, cost about $46, but in the initial year the prices were heavily subsidised so that the buyer paid only about $6. In a follow-up survey, 93.5% of the people interviewed expressed themselves satisfied with their stoves.

In 1982, the contribution from the family was increased to $9, and the shelter was omitted. Some 315 stoves were built during the year. The programme has suffered from poor follow-up, with the result that a number of women have been using their stoves improperly and not saving much fuel. Others have abandoned the stove altogether.

In West Africa, women usually cook sitting on a low stool. This traditional "stove" in Niger is made from scrap metal.

India

The Safai Vidyalaya (Sanitation Institute), which began work on improved stoves in Gujarat in the 1960s, is now promoting a two-hole stove with a chimney. It is made with burnt bricks, with a precast concrete slab with potholes for the top. There is a baffle under the second cooking hole and a damper between the two potholes. By mid-1983, the Institute had installed 35,000 stoves in Gujarat, making this by far the largest programme in India. Considerable alterations and modifications, often detrimental to the stove's performance, are made by the people who own them. An overall success rate of about 50% is claimed for the programme.

"No systematic evaluation survey has been carried out and these figures are based on random and informally collected feedback. Also, given the organisation's primary concern with sanitation, the main emphasis has been on smoke removal, and not energy conservation. The rating of a stove as successful seems to be largely based on its being used continuously, with or without modifications", according to a 1983 report on the programme.

The Tata Energy Research Institute (TERI) near Pondicherry has been working to improve the local two-pot *chula*. This programme is based on the conviction of the director, Chaman Gupta, that projects relying upon the introduction of stove designs evolved in a laboratory have little chance of success. TERI is trying to gradually improve existing local designs, in the hope that this will lead to increasingly sophisticated stove types.

The Thapar Polytechnic has been working in the Punjab to introduce a modified version of the *HERL chula* developed in the 1950s. One village has been saturated with 100 of these stoves. Their real cost is the equivalent of US$1, but they are sold for only 20 cents. The Centre for the Application of Science and Technology to Rural Areas (ASTRA), at the Indian Institute of Science in Bangalore, also has a stove programme.

An Indo-German project in Himachal Pradesh is introducing a new stove developed from local trials. Called the *Dhauladhar chula*, it is made of mud, can have two, three or four potholes, and incorporates dampers and a chimney. Dissemination started in 1981, and by mid-1983, 950 stoves had been built in the project area, and a further 550 outside it.

The *Nada chula* was developed by Madhu Sarin while she was working on a Ford Foundation development project in villages near Chandighar. This does not rely on a closely defined stove design, but is instead a system for stove design. Stoves are made of mud and incorporate a draught control door in front of the firebox as well as a chimney damper. The size and number of potholes and the overall size and location of the stove are decided in accordance with each user's needs and preferences.

Dissemination relies on women learning how to make the stove and then building stoves in the nearby areas. Many prejudices and difficulties are encountered in drawing women into such activities, but around 100-200 stoves have been built. The number of trained women builders is increasing and the project area is gradually being extended.

Other countries

Stove initiatives are taking place or are being developed in such African nations as The Gambia, Liberia, Lesotho, Burundi and Mali. In Asia, efforts are being made in Fiji, Bangladesh, Papua New Guinea and other countries.

Guatemalan technicians have helped to launch *Lorena* stove programmes in Jamaica and the Dominican Republic. These stoves have also been introduced to Mexico, Honduras and Ecuador, where Peace Corps volunteers have helped to build about 200 *Lorenas*. In most of these programmes the number of stoves built has been small. In Nicaragua, the government is engaged in some initial stove design and development work.

Chapter 7

But do they save wood?

There is no doubt that "improved" stoves can be thermodynamically more efficient than traditional stoves and open fires, when tested under controlled conditions. The evidence for their improved efficiency in the laboratory is not in dispute.

The question is whether stoves which save fuel in the laboratory can save fuel in the villages and cities of the Third World. Little data is available. Many stove programmes are operating in virtually total ignorance of whether or not they are meeting their fuel saving objectives. The position was described by Timothy Wood, writing about the Sahel, as follows:

> "We still do not know how much wood is saved by using 'improved woodstoves' . . . One often hears about the total number of stoves built by this or that project. It would be more useful to know how much wood is actually being saved as a result of these stoves. Surprisingly, most project managers do not know, and many do not seem very interested."

Few follow-up studies have been carried out. Few of these meet the requirements of valid statistical analysis. Some simply rely on hearsay, or asking people if their new stoves save energy; both methods are notoriously unreliable guides to what is actually happening. Such reports, without quantitative evidence to confirm them, cannot be used as a basis for serious estimates of overall fuel savings.

Fuel consumption surveys require a certain minimum number of observations if there is to be any confidence that the results are not purely random or due to a chance combination of factors. One of the measures used by statisticians is defined as the "level of significance" of the results of a particular set of observations: a 5% level of significance gives a good assurance that the results are not simply a matter of chance; a 10% level of significance is still reasonable.

In the Provisional International Standards for Stove Testing agreed among stove experts in December 1982, a table is given of the sample size necessary to assure a 5% and 10% level of significance from test results showing varying levels of expected fuel savings (see table). Results based on numbers smaller than these are less reliable the further they fall short of the required sample size.

Percentage Fuel Savings Expected	10% Level of Significance	5% Level of Significance
	- minimum number of households -	
10%	54	92
20%	14	23
30%	7	11
40%	5	7

Source: Vita (1983)

Necessary sample size for fuel consumption surveys

Results of fuel consumption surveys

One of the few reported sets of fuel consumption tests from Latin America was carried out in Guatemala and reported by Susan Bogach in 1981. Seventeen families were chosen in the village of Tecpan in the vicinity of Guatemala City, and surveyors monitored their consumption of fuelwood over a two-week period, during which they cooked normally over a three-stone fire.

Experimental models of five different improved stove models were then installed. One was a substantially modified *Lorena* stove with three potholes in line and a chimney. Two different versions of a three-pot brick stove with metal cooking plates and a chimney were also used, together with a two-pot version of the same type. All were fitted with draught controls in the form of butterfly valves in the chimney. In addition, one of the stoves used locally, which has a large firebox, a metal cooking plate and no draught control, was used.

The owners were given a week's training in the use of their new stoves. Their fuel consumption was then measured over the following two weeks. Each stove saved fuel compared to an open fire. But there was considerable variation in the savings, depending on the type of stove. The least savings (13%) were obtained with the traditional stove. The more elaborate stoves with draft controls gave savings of 27-49%. The *Lorena*-type showed a saving of 36%.

For each stove type, the sample size falls well short of that required for statistical validity, though the average saving over the whole group clearly indicates a beneficial effect from using the stoves. However, the report mentions that in follow-up visits it was found that in some cases the damper and baffle controls were opened fully to light the fire and then left half-open for the rest of the day, rather than being carefully used to regulate the rate at which the fire burned. It is not stated what, if any, effect this had on fuel consumption.

In March 1980, fuel consumption tests on six *Lorena*-type stoves were carried out in Sri Lanka. The results showed these stoves to be at the lower end of the efficiency range of traditional fireplaces measured in the field tests. Preliminary results from the introduction of an enclosed stove based on the *Lorena* design into the Kandy area in Sri Lanka showed a similar disappointing performance. The stoves were not markedly better than a well-tended open fire, and some poorly constructed models performed far worse than the traditional fire. Both programmes were changed radically.

Jean Bernard Roggeman has carried out a fuel consumption survey of about 100 families in Upper Volta and Senegal, the Senegal reference group being in a village which had had no contact with stove programmes and where the open fire or metal tripod were used. The second group was composed of users of a banco stove without a chimney, and the third group used a banco stove with a chimney. Both the latter groups were in villages in the same region and comparable with that of the reference group.

In Upper Volta, the reference group was in a poor quarter of Ouagadougou where the three-stone fire is the normal cooking method; a second reference group was chosen in a slightly more affluent quarter. Two groups of stove users were selected, one using the *Nouna* stove, the other the *Kaya*.

Families were observed while cooking a typical meal and the quantities of wood used were measured. Roggeman sought comparability between the different groups in income and type of food cooked. But there were considerable differences in the types of fuel used, the meals cooked and the sizes of the families for which the meals were prepared.

The results showed a consumption range of 2.4-3.0 kg (5.3-6.6 lb) per meal per family for those using the three-stone fire or metal tripod. The average consumption for the banco stove with chimney was 1.8-2.6 kg (4-5.7 lb) per meal. The consumption range for the banco stove without a chimney was 1.6-2.4 kg (3.5-5.3 lb) per meal. The *Nouna* stove used 3.9 kg (8.6 lb) per meal, and the *Kaya* 4.0 kg (8.8 lb).

At the family level, this represents a saving of 20% for the banco stove with a chimney, and 30% for the banco stove without a chimney over the figures obtained for the open fire and tripod. Both the *Nouna* and the *Kaya* stoves consumed considerably more than the open fire. But these results are difficult to interpret because the family sizes differed among the various groups.

Full details of all the family sizes are not given in the report, but recalculating on a per head basis for those that are given radically changes the fuel economy ratings. The average consumption on the open fire and the tripod is the same as that for the banco stove without a chimney. The banco stove with a chimney, the *Kaya*, and *Louga* stoves all have a per head consumption about 33% higher.

Later surveys in Senegal by Gerard Madon relied on measuring the fuel consumption before and after the installation of different types of *Ban ak Suuf* stoves. In several areas where such stoves were about to be installed, about five families were selected in advance. Over a period of one week their consumption of wood or charcoal was measured. Detailed records were kept of one family called the "pilot family", which partly played the role of a control group in the experiment. A total of 87 families in 19 different locations were chosen. After the stoves had been installed, the families' fuel consumption was measured for a further week.

Most of the families saved fuel, the average savings being 26%. But the range of variation was quite high, even within the individual villages. In the village of Bandegne, one family showed a saving of 35%, while another showed an increase in consumption of 25%. In the village of Tivaoune, the change in consumption varied between a saving of 52% and an increase of 16%.

Most of the families which used more fuel were using cooking pots which were too small for the potholes. Stoves with chimneys saved more than stoves without chimneys.

The savings in the pilot families were about 35%, as opposed to 22.5% for the remainder. As these pilot families received most attention from the survey team, it adds weight to the suspicion voiced by Timothy Wood that people may feel indebted to the promoters of a stove project, and try to economise on their fuel consumption in return.

In Upper Volta, some people using improved stoves of the *Nouna* type were interviewed in 1980, and all claimed savings of between 30% and 50%. To test this, several families were selected with the intention of monitoring their fuel consumption for one week using an open fire and for a second week using a stove. Due to the practical difficulties of carrying out the tests, comparative figures were only obtained for two families. These results, which extended over three days in one case and four days in the other, failed to prove any fuel savings at all. One family was found to use an average of 35.8 kg (79 lb) of wood per day on the open fire, and 34.4 kg (75.8 lb) per day on the stove. The other used 3.9 kg (8.6 lb) per day on the fire and 4.5 kg (9.9 lb) on the stove.

Also in Upper Volta, observations reported by Gautam Dutt compared the per capita fuel consumption in preparing the evening meal in 10 families. Five were using an open fire, and five were using a *Kaya*-type stove in tests over a four-day period. Although the families were not comparable in the food they cooked, the sizes and numbers of pots they used, and their detailed numbers and patterns of behaviour, a fuel saving of about 37% was recorded for the stoves. In some other tests carried out in Niamey, comparisons were made between per head daily consumption on 31 *Kaya* stoves and 26 users of the traditional *foyer malgache*. The *Kaya* stoves showed a saving of approximately 14% in comparison with the traditional stoves.

In another series of tests carried out in Ouagadougou, two sets of poor families were selected; one group was given new stoves. The wood consumption of each family using the traditional open fire was monitored for a week. Meanwhile, stoves were built in the homes of the other families, the household cook was taught how to use the new stove and encouraged to use it for two weeks. The stoves were of the *GS* type developed for use in Upper Volta: a two-hole sand and clay model with a chimney.

Twenty-one families were enrolled in the project, but four dropped out. The number of people served at meals ranged from four to 20, with an average of 10.4. The average per capita consumption of all the families was 0.93 kg (2 lb) per day. Seven families were given stoves and their consumption was 0.41 kg (0.9 lb) per head in the third week, a saving of 55%. When the families without stoves were re-surveyed, they showed a consumption of 0.84 kg (1.8 lb) per head, a reduction of 10%.

In Nepal, a follow-up study was done on a 10% sample of the 690 *insert*-type stoves installed in the Kathmandu valley some six to eight months previously. Surveyors found that 68% of the people were using the stove daily or frequently, 15% were using it infrequently and 18% not at all.

Of those interviewed, 82% felt that the stoves used less fuel and cut cooking times. The mean fuel savings were estimated at 38%, but there are no details of how this figure was established. The report notes that laboratory tests on the *insert* stove gave savings of 36%.

Tests were carried out in 1982 on 21 improved stoves in Indonesia. These were boiling water tests, allowing a calculation of the "percentage of heat used" (PHU), a measure of thermal efficiency. Nine stoves of the *Lowon* type (a two-pot stove without a chimney) and seven of the *Katesan* type (three-pot with chimney) were tested and compared with traditional stoves. The owners of the *Lowon* stoves had said that they saved about 50% of their previous fuel consumption through the use of the stove.

The tests found the average percentage of heat effectively utilised in the *Lowon* stoves to be 16.5%, in comparison with 14.4% for the traditional stoves. The performance of the *Katesan* stoves, measured as the percentage of heat utilised, was virtually identical with that of the traditional stoves.

In Fiji, typical Fijian, Indian, and Chinese meals were cooked on an open fire and a stove. The stove was a four-pot design made of concrete which is being promoted in rural areas. In all cases the open fire used less fuel than the stove. For the Chinese meal it used between 21% and 47% of the fuel required for the stove, and for the other meals it used 51-67%. But the same type of stove being used in a village consumed 1.28 kg (2.82 lb) of wood per head per day, compared with 1.57 kg (3.46 lb) per head for open fires, a saving of 35%.

A set of boiling tests in Fiji reversed this pattern, showing the stove to have an efficiency in the range 6.1-10.5%, compared with 3.8-5.1% for the open fire. Field observations, however, showed that the stove used about as much wood as an open fire.

Little is known about the longer term performance of stove users. Most consumption surveys have been carried out immediately or soon after the stoves have been installed. The new owners are self-conscious about fuel use and aware of being under observation. They are therefore unlikely to waste fuel. Timothy Wood made the following observation on surveys conducted when the stove is new:

> "Using the stove is still a novelty, the user gets daily attention and is motivated to conserve wood. In Upper Volta, two weeks of daily measurements of wood consumption in households *without* stoves (author's emphasis) were alone sufficient to reduce consumption by 25%".

The stoves deteriorate, and badly damaged stoves use energy less efficiently than new stoves working properly. Studies are needed on how well different stoves last in practice. The number of people continuing to use them is also a major factor in determining the overall savings from any particular programme. There is little such information available.

It is important to distinguish between mud stoves and those made of more durable materials. These latter also tend to be more accurately made, and when they save fuel in the beginning there is a better chance that they will continue to do so. Nevertheless, they too deteriorate with time. Only long-term studies, which have yet to be carried out, will determine how well their performance is maintained. In the case of stoves with chimneys, failure to clean the stove and flue properly and to maintain the draught controls in working order can also lead to malfunctioning and an increase in fuel consumption.

It is obvious that a great deal remains to be learned about how much fuel is actually saved by improved woodstoves. Given the limited data available, one can draw almost any conclusion one prefers. The evidence is unsatisfactory from a statistical point of view and is often contradictory.

The best that can be said with assurance is that some stoves save fuel initially; some do not, and little is known about what actually happens in most cases, particularly in the longer term. But there is no justification for the often repeated claim that stoves save 50% of the fuel normally used.

Chapter 8

Improving improved stove programmes

A great deal of time, money and effort has been invested in designing and implementing stove programmes. Through this work it is becoming clear that a dissemination strategy which is suitable for one area will not necessarily work in another. It is also clear that bringing a stove from the laboratory to widespread practical use is a complex process.

The number of stove users required if a programme is to make any fuel-saving or social impact is far too great for any external agency to construct and supply every stove. Self-sustaining dissemination must therefore be the ultimate goal of any stove programme.

At a workshop held in Sri Lanka in April 1981, a nine-step approach to dissemination was agreed among the stove programme directors present. These steps were:
1. An initial needs-assessment study;
2. Development of stove design criteria based on the study;
3. Modification of existing stoves or design of new ones;
4. Laboratory testing of the stoves;
5. Field programme to test performance and acceptability of the new stoves;
6. Monitoring and evaluation of the field programme;
7. Further design work if new stove is not satisfactory;
8. Development and appraisal of methods of manufacture;
9. Design and implementation of large-scale dissemination programme.

Many programmes have failed because they omitted some of the above steps, or took them in the wrong order. This is particularly true of the initial design stages. In some cases, design criteria have been drawn up in the laboratory, with no reference to the actual needs of stove users. In other cases, dissemination efforts have begun before stoves were adequately tested in the field. This left field workers promoting stoves which quickly turned out to be faulty or unsatisfactory.

Commercially viable stoves

Where traditional stoves are already on sale and widely used, the introduction of new stove designs can be a straightforward process. It will be based on a local manufacturing and distribution infrastructure which already exists. To become part of this system, a new stove will have to have attractions, whether in fuel economy or otherwise, which will make people want to buy it. Unless it is to be imported from outside the area, local stove-makers must be able to manufacture it, and to make an adequate profit from doing so.

An external agency can introduce an improved stove into such a system by the following steps:

1. Design and test performance of a stove which appears suitable for local use, can be manufactured locally, and has a reasonable possibility of being economically viable;
2. Carry out marketing tests in which selected users are given stoves on a trial basis;
3. Study the performance and acceptability of the stoves;
4. Revise the stove design in the light of the findings, and retest if necessary.

The next step is to investigate local manufacturing capabilities in detail and select partners for the first phase of the dissemination programme. The design may need to be modified again in order to help stove-makers make the best use of their time, materials and technical resources. But design alterations made to suit local manufacturing methods should not hurt the performance of the stove.

Financial assistance will probably be needed in order to set up local manufacturing on a pilot project basis. After a further trial period, the project should be reviewed and changed if necessary. It should then be possible to proceed to large-scale publicity to encourage both manufacturers and customers to shift to the new stove.

Once a self-sustaining manufacturing and distribution system made up of autonomous stove-makers and sellers has been established, the role of the promoters is virtually over. Except in the initial phases, when it is necessary to stimulate customer interest, direct subsidies to buyers or manufacturers should have no part in the dissemination strategy of this type of programme.

Low-cost or no-cost stoves

Other programmes may be based on trying to disseminate improved stoves in an area where the market for stoves is weak or non-existent. This is the case in much of the Sahel, where open fires or extremely cheap tripods are used, in Asia, where home-made mud stoves are used, and in much of the rural Third World. In these circumstances there is rarely a commercial market for domestic fuels.

As a result, stove introduction and dissemination tend to be much more difficult. The commercial incentive for consumers to invest in stoves is virtually non-existent; nor are there locally available stove-makers. Programmes have to start almost from scratch in generating consumer awareness of the existence and benefits of new stoves, and then in creating manufacturing and distribution capabilities.

One approach in these circumstances is to endeavour to design and disseminate a no-cost stove — no-cost, that is, to the users and built of free materials with free labour. It will not, of course, be without cost; the stove programme will have to carry the full design, promotion, and extension costs.

But the no-cost stove has a number of serious problems. It must rely on materials which are available at no cost, close to the dwelling in which it is to be built. This makes it difficult to exercise quality control over the construction materials, whether banco, lorena or mud. In addition, construction must be done by the householder, who will not be a trained stove builder.

No-cost stoves must therefore be simple and easy to build. Their performance must not depend on construction to strict specifications. They will deteriorate rapidly. This means they will save less energy, over a shorter time, than will professionally made stoves sold in the market. The incentive to use them will be lower. So also will the justification for investing resources in programme promotion.

The use of direct subsidies

All stove programmes involve some degree of subsidy; the costs of promotion and setting up training and prototype programmes have to be borne by external agencies. But most programmes want to produce a stove which will either be bought by people at an economic price, or be built by the users themselves.

The question of more substantial subsidies occurs where an improved stove is unlikely ever to be economically viable in its own right. Essentially this means that the promoting agency must be prepared to play a welfare role, providing people with a stove which they would not be prepared to buy at its true commercial price. There will be no prospect of the programme becoming self-sufficient, and its impact will be decided by the funds at the disposal of the promoting agency.

Or subsidies may be justified on the grounds that market conditions will change, making the stove economically competitive at some time in the future. By making preparations in advance, by creating consumer awareness and a manufacturing and distribution system, the intention is that the stove will be available for entry into the market when it becomes commercially viable.

Setting an appropriate level for a subsidy is difficult. The temptation is to use a large subsidy to achieve rapid sales. But this prevents independent local entrepreneurs from developing competing versions of the improved stove, and reduces the prospects for self-sustained local dissemination.

Ultimately the objective must be to enable the stove to become commercially viable so that the subsidy can be phased out. An approach being tried in Nepal shows one way of phasing out subsidies. Ceramic inserts for mud stoves are given to users free of charge, in the hope that they will purchase replacements when the time comes.

The "half-life" concept

Timothy Wood has come up with a way of calculating the impact of a stove programme which uses the concept of "half-life" from radiation physics. The half-life of a radioactive isotope is the length of time taken for its radiation to fall to half its original level. The decay of each individual atom is not predictable, but that of the total number is, and each material has its own distinctive half-life.

Wood suggests that the same idea might be applied to new stove models. Each individual stove will deteriorate at a rate which is determined by the skill of its builder, the extent to which it is used and the environment in which it is placed. The half-life would therefore be the time it takes half a given number of stoves to fall into disrepair.

Experience in West Africa showed that the stoves being promoted had a half-life of between one and two years. If the output of stoves from such a project is constant, say 1,000 per year, and the half-life is one year, the maximum number of stoves which will ever be in use is 1,999. No matter how many years this 1,000-stoves-per-year output remains constant, there will never be more than 1,999 stoves in use. Even if the half-life is quadrupled to four years, the maximum number will be only 6,276.

The calculation shows that the only way in which a large-scale diffusion of stoves can take place is by a continually expanding programme. This means that manufacturing must be taken up by local stove-makers on a self-sustaining basis. It further reinforces the case against using any form of subsidy on anything but a temporary basis.

Justifying programmes

Stove programmes do not always consider the basic question of whether the investment of external funds and the consumption of scarce local resources to further the cause of improved stoves is justified.

Promoters need to assess the benefits to users of any given stove programme and match these against the investment necessary to obtain them. If the long-term goal of self-sustaining and spontaneous dissemination is to occur, there must be a reasonable match between the costs and benefits of the programme, particularly as these are perceived at a local level. This dilemma has been summed up by David French, working in Malawi. He was writing specifically about the promotion of no-cost mud stoves, where the prospects of long-term energy savings are the poorest:

"If significant wood savings are not a serious prospect, stoves have to be considered more carefully alongside the whole range of ways in which domestic life can be improved: e.g. by training in health or nutrition, inclusion of women in agricultural extension programmes, provision of potable water, improved schools and clinics etc, etc. When stoves are viewed in this way — in relation to the alternative ways of helping women — both women and governments are almost certain to assign them a very low priority."

This is not to say that no mud stove programme can ever be justified. The success and popularity of the simple *Louga* stove in Senegal would appear to be a case of a programme which is already succeeding, and could possibly continue under its own momentum if external support were withdrawn. Nor is it entirely fair to suggest that resources going to stove programmes are necessarily available for other projects.

A stove promotion programme may not be the absolute optimum investment of resources; but then few projects of any kind can stand this kind of test. The basic requirement is that a stove programme should be able to provide a rational justification for itself and for the money invested in it. Only those which can do so, and which show signs of ultimately becoming independent and self-sustaining, can warrant the investment of effort and resources involved.

Further reading

The following is a selected list of background reading material. In most cases, the organisations listed have continuing stove research and implementation programmes and can be contacted for their latest reports and information.

BOILING POINT. Regular bulletin of the ITDG Stoves Project, ITDG, London.

COOKSTOVE NEWS. Regular bulletin of the Aprovecho Institute, Eugene, Oregon, US.

DECHAMBRE, Gilles. *Le Développement des Foyers Améliorés au Sahel*, Association Bois de Feu, Marseille. 1982.

EVANS, I. and M. Boutette. *Lorena Stoves*. A Volunteers in Asia publication, Aprovecho Institute, Eugene, Oregon. 1981.

FAO. *Map of the Fuelwood Situation in the Developing Countries.* FAO, Rome. 1981.

FLAMME. *Etudes, Recherches, et Actions pour l'Economie des Energies Domestiques.* CILSS, Ouagadougou, Upper Volta.

GERMAN APPROPRIATE TECHNOLOGY EXCHANGE. *Helping People in Poor Countries Develop Fuel-Saving Cookstoves.* Study by the Aprovecho Institute for Deutsche Gesellschaft für technische Zusammenarbeit (GTZ), Eschborn, West Germany. 1980.

GOULD, H. and S. Joseph. *Designing Stoves for Third World Countries.* ITDG, London. 1978.

JOSEPH, S. et al. *Compendium of Tested Stove Designs.* Prepared for the FAO by ITDG, London. 1980.

KAUFMAN, M. *From Lorena to a Mountain of Fire.* Yayasan Dian Desa, P.O. Box 19, Bulaksumar, Yogyakarta, Indonesia. 1983.

DE LEPELAIRE, G., K.K. Prasad and P. Verhart. *A Woodstove Compendium*. Prepared for the Technical Panel of Fuelwood and Charcoal, UNERG, Nairobi. 1981.

MADON, G. and M. Gueye. *Programme de Diffusion des Cuisinières 'Ban-ak-Suuf'*. Rapport d'Activite no 4, April 1982, CERER, Dakar, Senegal. 1982.

MANIBOG, F.R. *An Interim Assessment of Improved Woodstove Components in Bank Fuelwood, Forestry and Rural Development Projects*. Energy Dept., World Bank, Washington, DC. 1982.

MICUTA, W. *Modern Stoves for All*. Bellerive Foundation, Geneva. 1981.

PRASAD, K.K. *A Study on the Performance of Two Metal Stoves*. Woodburning Stove Group, Technical University of Eindhoven and TNO, Apeldoorn, Netherlands. 1981a.

PRASAD, K.K. (ed.). *Some Studies on Open Fires, Shielded Fires and Heavy Stoves*. The Woodburning Stove Group, Eindhoven University of Technology and TNO, Apeldoorn, Netherlands. 1981b.

PRASAD, K.K. *Woodburning Stoves: Their Technology, Deployment and Economics*. Prepared for Technology and Employment Branch, Employment and Development Office, International Labour Office, Geneva. 1983.

ROGGEMAN, J.B. *Les Fourneaux Améliorés dans le Sahel – Rapport sur les Characteristiques thermiques des fourneaux améliorés à Bois*. Club du Sahel, CILSS, Ouagadougou, Upper Volta. 1980.

SARIN, M. *Chula Album*. 48, Sector 4, Chandigarah 160 001, India, and Ford Foundation, New Delhi. 1981.

VITA/ITDG. *Wood-Conserving Cookstoves, A Design Guide*. VITA, Mt Ranier, Maryland. 1980.

VITA. *Testing the Efficiency of Wood Burning Cookstoves: Provisional International Standards*. VITA, Mt Ranier, Maryland. 1982.

WOOD, T.S. *Improved Woodstoves in the Sahel: A Critical Assessment*. Proceedings of Workshop on Energy, Forestry and Environment. USAID Africa Bureau. 1982.

For Product Safety Concerns and Information please contact our EU
representative GPSR@taylorandfrancis.com
Taylor & Francis Verlag GmbH, Kaufingerstraße 24, 80331 München, Germany

www.ingramcontent.com/pod-product-compliance
Ingram Content Group UK Ltd.
Pitfield, Milton Keynes, MK11 3LW, UK
UKHW021822240425
457818UK00006B/35